Vital Subjects

Race and Biopolitics in Italy, 1860–1920

Transnational Italian Cultures 1

Transnational Italian Cultures

Transnational Italian Cultures will publish the best research in the expanding field of postcolonial, global and transnational Italian studies and aim to set a new agenda for academic research on what constitutes Italian culture today. As such, it will move beyond the physical borders of the peninsula as well as identifying existing or evolving transnational presences within the nation in order to reflect the vibrant and complex make-up of today's global Italy. Privileging a cultural studies perspective with an emphasis on the analysis of textual production, the series focuses primarily on the contemporary context but will also include work on earlier periods informed by current postcolonial/transnational methodology.

Vital Subjects

Race and Biopolitics in Italy, 1860–1920

Rhiannon Noel Welch

LIVERPOOL UNIVERSITY PRESS

First published 2016 by
Liverpool University Press
4 Cambridge Street
Liverpool L69 7ZU

British Library Cataloguing-in-Publication data
A British Library CIP record is available

ISBN 978-1-78138-286-8 cased

Typeset by Carnegie Book Production, Lancaster
Printed and bound by CPI Group (UK) Ltd, Croydon CR0 4YY

Contents

To Calvin and Michelle Welch,
with infinite gratitude

List of illustrations

Acknowledgements

This book has been many years in the making, and is the fruit of numerous conversations and exchanges undertaken with cherished interlocutors from San Francisco, Berkeley, and Santa Cruz to Ithaca, NY; Lancaster, PA; New Brunswick, NJ; New York City, Rome, and many places in between. My series editors at Liverpool University Press, Derek Duncan and Emma Bond, were a delight to work with. I thank them for their enthusiasm and for their roles in including this book in their brilliantly formulated and timely new series. LUP's Anthony Cond and Patrick Condon Brereton were superbly attentive to all aspects of the book's production. Thanks also to Sue Barnes and the typesetters at Carnegie Book Production for their patience, skill, and attention to detail. Catherine Belloy at the Marian Goodman Gallery and Maria Murguia at the Artists Rights Society were key in securing permissions for the work by Giuseppe Penone reproduced on this cover.

This book began during formative years as a doctoral student at UC Berkeley, where many of its core ideas took shape under the generous guidance of Barbara Spackman, Mia Fuller, Albert Ascoli, and Eugene Irschick. Additionally, Saidiya Hartman was an influential and enthusiastic early reader. Donald Moore's course first introduced me to Michel Foucault's lectures on governmentality and biopolitics and got me thinking about their relevance to Italian racial discourse. An American Council of Learned Societies Andrew W. Mellon Dissertation Fellowship enabled me to conduct much of the research that has gone into this book. I conducted the bulk of this research during an unforgettable fall and winter at the McGill University Library in Montreal, where I was generously hosted by family away from home: Cristiana Giordano, Kelly McKinney, and Bruce and Beatrice Mellett.

A heartfelt thanks to my doctoral advisors Barbara Spackman and Mia Fuller for over fifteen years of care and attention to my development as a reader, thinker, and writer. Barbara's thoughtful readings of nearly every word of the manuscript were invaluable; as a mentor, she has been rigorous, honest, and truly inspiring. In addition to sharing her expertise on critical theoretical approaches to literature and culture and patiently guiding me through Gabriele D'Annunzio's challenging and immense oeuvre (Chapter Three), it was Barbara who first called my attention to the works of Paolo Mantegazza (Chapter Two). Mia has been a precious interlocutor throughout, sharing her expertise on Italian colonialism and racial culture(s) with wit, nuance, and generosity. I'm also indebted to Mia for first drawing my attention to Leopoldo Franchetti's writings on colonialism and the southern question, which are the topic of Chapter One. Both of these women have shaped the readings in this book in innumerable ways, and I am eternally grateful.

Timothy Campbell, Suzanne Stewart-Steinberg, and Elena Past provided attentive and productive close readings of the manuscript at various stages. In addition to their respective works, their collective insights into thinking about modern Italian culture through political theory have been indispensible. I am grateful to Tim for sharing his expertise on biopolitics, but also for suggesting Giovanni Pastrone's *Cabiria* (Chapter Four) as an interesting text to add to the manuscript. In addition, Tim's encouragement throughout the years has been fundamental. Suzanne also kindly put me in touch with Erica Moretti, to whom I am indebted for sharing some of the images in Chapter One, along with precious archival material and guidance on Franchetti and Hallgarten's schools in Umbria.

My years in the Cornell University Romance Studies Department and as a Society for the Humanities Mellon Postdoctoral Fellow were immensely productive, and friendships with colleagues played no small role in keeping my spirits high through many a dark Ithaca month. In addition to Tim, I thank Aloja Airewele, Kitrina Baxter, Tamar Carroll, Rebecca Givan, Cary Howie, Patricia Keller, Tracy McNulty, Alice Michtom, Steve Pond, and Lars Schumann for keeping me intellectually engaged and emotionally grounded. At Franklin & Marshall College, I benefitted from the support, encouragement, and camaraderie of Antonio Callari, Linda Hasanuma, Peter Jaros, Carrie Lanfried, Giovanna Faleschini-Lerner, Scott Lerner, Gayatri Menon, Gianluca Rizzo, and Brenna Stuart. Karen Bassi and Deanna Shemek, whose profound and playful course on Machiavelli I happily took as an undergraduate, were wonderfully supportive during my brief but memorable time in the Literature Department at UC Santa Cruz. I thank Roberto Esposito, Sandro Mezzadra, Cristina Lombardi-Diop and Caterina Romeo for their generous feedback on the book's introduction.

I am incredibly lucky to have found such a welcoming group of colleagues at Rutgers University. My colleagues in the Department of Italian have been unfailingly supportive. Thank you to Andrea Baldi, Paola Gambarota, Ilona Hrenko, David Marsh, Laura Sanguineti-White, Carmela Scala, and Alessandro Vettori for making Rutgers such a pleasant and productive intellectual home. A special thanks to Paola for her attentive reading of the manuscript, her sage advice and razor sharp insights, and for inspiring my confidence in the project when it faltered. Warm thanks also to Sheri La Macchia, an irreplaceable administrator who has brought a wealth of knowledge, collegiality, and a true spirit of collaboration to all of our departmental operations. Rudolph Bell has been a consistent point of reference since my arrival at Rutgers, and has provided valuable counsel on a great many things, including every aspect of this book's production—from kindly helping me parse anonymous reviewer feedback to guidance on navigating the world of academic publishing. I'm grateful also to my students and colleagues in Comparative Literature, History, the Center for Race and Ethnicity, Cinema Studies, and throughout the School of Arts and Sciences for making Rutgers such a rewarding place to teach and learn: Mia Bay, Anne Fabian, Sandy Flitterman-Lewis, Alberto Iozzia, Seth Koven, Susan Martin-Marquez, Fatima Nacqvi, Al Nigrin, Donata Panizza, Andrew Parker, Nicoletta Romano, Dianne Sadoff, Eleonora Sartoni, Meheli Sen, James Swenson, and all of my undergraduate and doctoral students in the Department of Italian and the Program in Cinema Studies.

Two other reasons my time at Rutgers has been such a joy: Marisa Fuentes and Rebecca Givan, both of whom were colleagues and dear friends well before I landed in New Brunswick, and with whom I've shared countless precious hours of commuter tips, professional and personal advice, eating, drinking, and laughter. One of the many pleasures of being an academic is the depth and endurance of friendships like these. I'm lucky to count among my dearest friends and colleagues in Italian studies Rebecca Falkoff, Aileen Feng, Cristiana Giordano, Stephanie Malia Hom, Ara Merjian, and Silvia Valisa, all of whom have nourished my intellectual and emotional growth in unquantifiable ways. Raffaella Di Castro has graciously hosted me on numerous research and writing stays in Rome, and has been an incredibly supportive friend throughout the nearly decade-long process of research and writing. *Grazie di cuore a tutti.* Among the most consistently generous of readers of this book in its various stages is Patty Keller, who is as brilliant as she is honest, compassionate, and humble. She has been a true and selfless friend throughout. My gratitude for her friendship and intellectual sustenance is boundless.

Last and certainly not least, I thank my oldest friends and family for their love and lightning-quick wit, and for helping me keep perspective by never letting me take myself too seriously. Lauren Adam, Rachelle Arenas, Laura Del Fava, Gabrielle Feuersinger, Jessica Raiter, Julia Reed, Ashley Richards, and Adriana Serrano have been my inspiring, kind, and hilarious support system for as long as I can remember. I cherish our friendships and look forward to the next thirty-something years. I thank my parents, Calvin and Michelle Welch, to whom this book is dedicated, for their unfaltering love, integrity, and guidance. They've shared their passion for language, a good story, and an even better sense of humor, which I consider among the greatest gifts. Their genuine commitment to community in all its finest forms is a daily inspiration. Prentice, Nathalie, Jalius, Jacob, and Lucas Welch have been a constant anchor, providing me with a warm sense of home even when I'm far from it. I'm grateful also to the Swords-Horrell family for welcoming me so graciously into theirs—Marti and Dana, Madeline, Peter and Diane, Alicia and Tim, Sol and Eileen, Andy and Sue, Lauren, Mary and Walter, Rolph, Susanne, Johannes, Mark, Sarah and Franklin, Miriam, Sam, and their immense extended family for all of their hospitality and care. I can think of no one with whom I'd rather face this world than Nathan Horrell, whose wisdom, warmth, solidarity, and humor sustain me every day. To you, deepest love and gratitude.

INTRODUCTION

Vital Subjects

> Italy is a country on the frontier, not only in a geographic sense, but also culturally, between different worlds, between Europe and the Mediterranean, between North and South [...]. Italy is traversed but also in a certain sense constituted by this fracture.
>
> Roberto Esposito (Campbell, "Interview with Roberto Esposito" 49)

In December 1887, Italian Prime Minister Francesco Crispi introduced unified Italy's first legislation on emigration with the following words:

> The Government cannot remain an indifferent or passive spectator to the destinies of [emigrants]. It must know exactly where they are going and what awaits them; it must accompany them with a vigilant and loving eye...it must never lose sight of them in their new home [...] to turn to its advantage the fruits of their labor. Colonies must be like arms, which the country extends far away in foreign districts to bring them within the orbit of its relations of labor and exchange; they must be like an enlargement of the boundaries of its action and its economic power. (Atti Parlamentari, 2a sessione AC 85)

Crispi was referring to what had become one of the central questions for policymakers after Italian unification: how to address the fact that millions of hard-working and newly nationalized Italians were leaving Italy, more and more often permanently, in search of better fortune in Europe and, in ever-increasing numbers, across the Atlantic (Fiore 71–82). In this period, colonies were considered both the "spontaneous" settlements of emigrants abroad and the planned settlements in East Africa for which, as early as 1887, Italians had been sent to fight in deadly battles.[1]

In describing the state's role in the regulation of emigration, Crispi stages a convergence between two modes of government. In the first of these two modes, government is a disciplinary agent, whose surveying (and "loving") eye is armed with knowledge and aimed at individual emigrant bodies. In the second, the aim of government shifts to include individuals as elements of a national population, whose borders and numbers must expand, enveloping new territories and reproducing itself, in order to survive. This second mode of government, known as biopolitics, was, in 1887, yet to be named as such, though European nation-states had long been operating under similar imperatives. It was not until the publication of *The State as a Living Organism* (*Staten som Lifsform*) in 1916 that the Swedish political scientist Rudolph Kjellén would draw from his earlier work on geopolitics in order to name "biopolitics" the role of the national population in the security of the state.[2] If geopolitics concerned the construction of the state through territory, biopolitics concerned its grounding in "the people." Kjellén writes: "One cannot divorce land from the state without the state as a concept losing its meaning, and the people leaving the land kills the state."[3] The state—as both territory and people—was thus itself an organic, and mortal, entity.

Prime Minister Crispi looked to African colonization (according to some accounts, Eritrea, Italy's first official African colony, was named by him in 1890) to resolve the fragmentation of laboring Italian bodies by implanting them on the—as yet missing—limbs of the colonial nation-state ("colonies must be like arms"). Crucially, these laboring bodies were themselves figured as little more than "arms." Agricultural laborers were named *braccianti* for the only tool they possessed, their *braccia* (arms). As Kjellén would put it,

1 For more on "spontaneous" colonies in the Americas and Crispi's demographic colonies, see: Choate, "From Territorial to Ethnographic Colonies and Back Again" 65–75.
2 On Kjellén's coining of biopolitics, see: Roberto Esposito's discussion of Kjellén's *Stormakterna: Konturer kring samtidens storpolitik* (1905) in Esposito, *Bíos. Biopolitics and Philosophy* 17. Kjellén also coined "geopolitics" as a politics that addresses a moment when "the great powers, vigorous states, endowed with a limited territory, discover the need for extending their borders through the conquest, fusion, and colonization of other lands." Kjellén's *Staten som Lifsform* has not, to my knowledge, yet been translated into English.
3 For an account in English of Kjellén's *Staten som Lifsform*, see Tunander. I cite her translation of Kjellén (457), though what Esposito translates as "biopolitics" Tunander translates as "ethnopolitics." Tunander is careful to rebuke post-World-War-II readings of Kjellén as a racist or a proto-Nazi (451).

"like man, the state may lose a limb without perishing, but 'there are [other limbs] without which the state could not survive'" (Tunander 453). In Crispi's formulation, the passage from individual to member of a national population occurs through labor: the collective "fruits" of individual laboring Italian bodies were to nourish the newborn nation-state. "The population" is thus born, so to speak, through labor. This rhetorical link between primarily agricultural labor productivity and biological reproductivity was not limited to debates about turn-of-the-century Italian colonialism and emigration. Instead, it represents a larger current in post-Unification racial discourse and constitutes the basis for a variety of questions that this book sets out to examine.[4]

Since World War II, Italy has struggled to recast both its colonial past and its alliance with Nazi Germany. For many years, pervading much intellectual and public discourse was the contention that, prior to the great influx of racialized migrants in the mid-1980s, and with the exception of the fascist "parenthesis," there simply was no race (racialized others, racist intolerance, etc.) in Italy. This book examines a selection of social scientific, political, literary, and cinematic texts from the years between Unification and the end of the World War I (*c.*1860 to 1920) in order to explore how race underpinned the discursive constitution of Italians as modern political subjects—a process often referred to as "making Italians," to quote Massimo D'Azeglio's ubiquitous but misquoted shorthand.[5] As a growing number of scholars are asking what the turn of the twentieth century in Italy can demonstrate about modernity and nation formation, this book contributes a reading of how race was integral to these processes, and not merely a marginal afterthought (Choate, *Emigrant Nation*; Gambarota; Stewart-Steinberg). To that end, anthropologist Miguel Mellino has recently argued:

4 Portions of this book have appeared in: Welch, "Razza e (ri)produttività. Per una lettura biopolitica della razza nell'Italia postunitaria e contemporanea." *L'Italia postcoloniale.* Ed. Cristina Lombardi-Diop and Caterina Romeo. Florence: Le Monnier, 2014, 79–91; "Race and Colonial (Re)productivity," *Annali d'italianistica* 32 (2014). *From Otium & Occupatio to Work & Labor in Italian Culture.* Norma Bouchard and Valerio Ferme, eds. 197–214.

5 In English, "We have made Italy, now we need to make Italians." D'Azeglio's dictum has become what Suzanne Stewart-Steinberg in *The Pinocchio Effect: On Making Italians (1860–1920)* refers to as an "almost mythic cliché" (1). The full statement (which appears to have been added by an editor as it does not appear in D'Azeglio's manuscript) in Italian reads, "Pur troppo, si è fatta l'Italia, ma non si fanno gl'Italiani" (D'Azeglio and Ghisalberti 5). For comparison, see: D'Azeglio, *I miei ricordi* 3. For a publication history of the dictum in question, see Hom 4–7.

> Taking as a necessary (postcolonial) starting point the underlying coloniality of [...] national formations [and] their necessary material inscription within the coloniality of modern global capitalist power means arguing that race, racism, and racialization have fractured the Italian space ever since the birth of the nation. (87)

Unlike scholars who have addressed race in this period through analytics of otherness, the stereotype, or the binary logic of inferiority/superiority, I read Italian formulations of race in a "vital" key, as I argue instead that a major point of articulation for Italian racial discourse is at the intersection of primarily agricultural labor productivity and biological reproduction. The present study thereby heeds Mellino's call to "de-provincialize" Italy by helping to "cast the shadow of race and racialization over the very act of foundation of the modern Italian nation" (88).[6] Four case studies—on liberal statesman Leopoldo Franchetti's (1847–1917) proto-sociological writings on the southern question and early colonialism, and his later pedagogical project at the Villa Montesca; Italy's first anthropologist Paolo Mantegazza's (1831–1910) moral-scientific treatises and novels; poet, novelist, and political provocateur Gabriele D'Annunzio's (1863–1938) decadent novels and speeches at Fiume; and early cinema pioneer Giovanni Pastrone's (1883–1959) imperialist epic *Cabiria*—allow me to focus on how distinct configurations of the rhetorical constellation of race and (re)productivity shaped post-Unification racial discourse, producing Italians as *vital subjects*. The readings of post-Unification cultural production that follow are also bound by their historical positioning vis-à-vis the convergence of the three most pressing post-Unification political anxieties about the (re)production and fragmentation of the Italian nation-state: the so-called southern question; mass emigration to the Americas; and early colonialism in the Horn of Africa and Libya. If the readings that follow are anchored in a particular time and place (Italy and its "missing limbs" between Unification and the end of World War I), on the other hand, they also interrogate both implicitly and explicitly some of the contemporary legacies and limitations of post-World-War-II anti-racism.

6 Mellino refigures Dipesh Chakrabarty's charge to provincialize Europe by suggesting we de-provincialize Italy by revaluating its role in the promulgation of capitalist modernity wherein both European-Christian humanist culture and race were necessary to the spread of colonial capitalism.

The Deputy, the Doctor, the Decadent and the Director

In the years between Unification and World War I, Italian racial discourse consistently transgressed disciplinary boundaries, shuttling across journalistic, literary, (social) scientific, and photographic media. This book deliberately engages texts produced across these fields in order to examine the pervasiveness of the biopolitical dimension of Italian racial discourse. As Lucia Re has argued with regard to Italian racial discourse and its literary intertexts in these years:

> There was effectively no border, no substantial difference between literary and scientific discourse, between fiction and poetry on one side and empirical reality and objective observation and description on the other. From the very start, the discourse of the human and social sciences and of positivist anthropology was saturated with fantasy; it absorbed and recycled literary images, literary devices, and fictional devices [...]. ("Italians and the Invention of Race" 18)

Similarly, Maria Sophia Quine has claimed that the historiography of the *Risorgimento* often depicts the "making" of the Italian nation-state as an exclusively aesthetic activity—confined to monuments, poetry, opera, and painting. Quine argues that social science should be "placed alongside the arts in a pantheon of patriotism," given that the founders of Italian anthropology (she names Giustiniano Nicolucci, Paolo Mantegazza, and Giuseppe Sergi) "all produced 'great' works of fantasy, 'fact' and 'fiction' about the nation" (151–152).

Each thinker explored in the coming chapters was a pioneer in some form of cultural production—sociological reportage, visual anthropology, public hygiene treatise, the bourgeois novel, or popular cinema—in post-Unification Italy, aimed at making bodies, be they laboring, sexualized, and/or racialized, legible.[7] Each transgressed conventional generic or disciplinary boundaries, between science and art, or between art and

7 In addition, three of the figures considered—Franchetti, Mantegazza, and D'Annunzio—served in the Italian Parliament during the period under consideration. The terms of each in public office overlapped: Mantegazza served as a parliamentary deputy (for Monza) from 1865 until 1876 and as a senator from 1876 until 1910. Franchetti served as a deputy (for Città di Castello) from 1882 until 1905 and was elected senator in 1909. D'Annunzio was elected a deputy (for Ortona del Mare) in 1897 and

politics, for example, in explicitly political projects that may be considered, at worst, flagrant propaganda and, at best, part of the patriotic guiding (or governing, in the Foucauldian sense) of national subjects. Franchetti was an influential career statesman dedicated to the agricultural education of the masses whose writings helped to bring the so-called southern question and "demographic" colonialism into the halls of Parliament. Mantegazza was a physiologist by training, who, after traveling the world armed with both monocle and notebook, became Italy's first chair of anthropology and went on to write countless popular treatises and romance novels on moral and physiological hygiene that enjoyed multiple translations and transnational circulation. D'Annunzio achieved international fame and infamy over the course of his literary and political career, publishing numerous volumes of critically acclaimed poetry, plays, and prose. Before his condemnation by postwar critics who found in his work the seeds of Mussolinian rhetoric and politics, D'Annunzio was celebrated in Italy for his unprecedented literary modernity. Pastrone first gained fame on a world stage with his colossal early film *Cabiria* (1914), which is credited with groundbreaking techniques such as the *carrellata* (or tracking shot, which was for years referred to in the U.S. as the "*Cabiria* shot").[8] More importantly, each of the works discussed here—from fiction to ostensibly non-fiction and film—stages the encounter between race and (re)productivity through a rhetoric of defense from collective contagion and degradation. This immunizing rhetoric was operative during this period in large part because it responded to a particular set of biopolitical anxieties about the implications of mass emigration and the potential presented by colonial acquisition. Chapter One of this volume discusses "Colonial (Re)productivity." At stake for Franchetti is the stability (read racial integrity) of the liberal-democratic state, threatened from within by the destitution of primarily southern farmers who stand to become agriculturally productive and sexually reproductive in lands, particularly the Americas, that fall outside Italy's juridical and ideological domain. Crucial to Franchetti's scheme to resolve the southern question through the "demographic" colonization of Eritrea is the management of Italian peasant

in 1900 made his famously theatrical transition from the right majority to the left; after suffering electoral defeat in that same year, he left public office.

8 Pastrone was also responsible for carving out a new role for the director-producer in an early Italian film industry that was gaining increasing recognition (during its "golden years" between 1908 and World War I) for its innovations in both production and publicity (Alvosio, "The 'Pastrone System'").

bodies, in terms of their territorial positioning (laboring bodies are rerouted to the Eritrean colony), daily alimentary intake, and biological and labor output.[9] The chapter discusses Franchetti's founding of an internationally renowned teacher training institute and school for peasants at the Villa Montesca in 1901 together with his earlier projects in southern Italy and Eritrea in order to illustrate the breadth of his colonial biopolitics as a means of making Italians.

In Chapter Two, "Immunitary Technologies," Paolo Mantegazza is haunted by the degenerate and diseased newborn, from his earlier, non-reproductive novel *Un giorno a Madera* (A Day in Madeira, 1868) to his later science-fiction novel about regulating reproduction, *L'Anno 3000* (The Year 3000, 1897). The first novel blocks the reproductive union of a tuberculoid Emma and her beloved William because of the threat posed to Emma's offspring by her hereditary illness. The later text explores the possibility of the birth of such a being, resolving the threat it poses to the body politic through an immunizing death sentence. In *L'Anno 3000*, material technologies of racial visualization (an x-ray aimed at rendering skin "as transparent as glass," a psychoscope that makes cerebral defects legible, and an incinerator into which biologically unfit newborns are tossed) are firmly rooted in *fin-de-siècle* fantasies about preserving the race, which is figured in this novel as emblazoned simultaneously upon and beneath the skin. Mantegazza's use of technologies of racial visualization in *L'Anno 3000* can best be understood in relation to his pioneering contribution to the field of visual anthropology through the use of photography in ethnographic research.

The third chapter, "Mutilated Limbs," draws from a selection of works across Gabriele D'Annunzio's vast production in order to show how his deployment of blood shifts from his early "decadent" novels—the *Romanzi della rosa* trilogy, made up of *Il Piacere*, *L'Innocente*, and *Trionfo della morte* and published between 1889 and 1894—to his writings and speeches at Fiume (1919–1921).[10] If in this first set of texts blood is confined to (tainted)

9 Franchetti's work does not attempt to provide a thorough account of colonial reproductive politics, though the Eritrean memoir of military wife Rosalia Pianavia Vivaldi published in 1901 addresses the abandoned children of unions between "white" colonials and "blacks" in a chapter curiously entitled "My Children." For a reading of the maternal rhetoric of Pianavia Vivaldi's text, see Lombardi-Diop, "Madre della nazione." For historical accounts of reproduction among Italians and Eritreans, see: Barrera, "Colonial Affairs"; Sòrgoni, *Parole e corpi*.

10 In English, the *Romanzi della rosa* are titled: *The Child of Pleasure*, *The Victim*, and *The Triumph of Death*.

genealogy, at Fiume the defense of the bio-territorial *patria* is enacted through the shedding of sacralized blood. Central to D'Annunzio's earlier novels is the thwarted heredity (both retrospective and prospective) of his protagonists: the bodies of each of his male protagonists are therefore emphatically not reproductive. Instead, two of his protagonists destroy life, one through murder and the other through infanticide. At Fiume, blood is shed by mutilated soldier-patriots in order to restore a specifically racialized Italian primacy in the Adriatic. Furthermore, a dismembered body is mobilized in Fiuman rhetoric to mirror and remedy the fractured territorial gains of Italy's "vittoria mutilata" (mutilated victory) in World War I.

How do these biopolitical logics of racialization migrate into the visual language of early cinema? Chapter Four, "Biopolitics and Colonial Drive," returns to the colonial frontier and to the question of visual technologies of racialization by way of Italy's first international blockbuster film *Cabiria*, which was produced to garner support for the Italian invasion of Libya (1911–1912). Yet the ambiguous racialization of the film's protagonist, Maciste, whose skin was visibly darkened for his role as the slave-hero, fails to conform to what one might anticipate from an imperialist epic (logics of absolute racial difference, inferiority, or otherness). Instead, in Maciste's artificial darkening I read a biopolitical racial logic, one that is haunted by Italian histories of "proletarian" emigration to the Americas (a term canonized by poet Giovanni Pascoli in his famed 1911 address *La grande proletaria si è mossa* (The Great Proletarian, She Has Risen!)). This biopolitical reading of Maciste's racialization rests on the fact that the film posits a territorial loss (former Roman territories in Mediterranean Africa) alongside a corporeal one (the loss of a noble Roman girl to Carthaginian hands) in order to present colonial war as paradoxically productive, or life-affirming.

Vital Subjects asks how the contours of Italian modernity were sketched in positive biopolitical terms that focused on productivity, vitality, and preservation. These terms, including but not limited to race, stock, and people, are imbued with an immunitary logic that folds in upon itself as it forecasts the eugenic "removal" and/or "suppression" of other, threatening forms of life. Significantly, these other forms of life are not necessarily members of an apparently external collectivity (the colonized populations of Eritrea or Libya come to mind as examples), but instead members of the however tenuously defined Italian community that such terms (race, stock, people) attempted to solder together in the first place. In the texts under consideration, these terms are rooted most deeply not in Rome, nor in the *Risorgimento* hotbed of Piedmont, but in the lands that constitute the "missing limbs" of the territorial *patria*: Eritrea, Fiume, Libya, and beyond.[11] For Franchetti, landless southern peasants who were denied political and economic sustenance in Italy were to find it in Eritrea; their implantation on Eritrean soil would also enable them to emerge from their marginalization to become "true" representatives of the *razza italiana* (Italian race). The implications of such claims are much more nuanced and far-reaching than can be explained with recourse to dismissive categorization as "racist" or "pseudo-scientific." The tenor of the post-Unification racial discourse under consideration was overwhelmingly positive, focusing on making Italians *vital subjects*—robust, vigorous, well-nourished, and (re)productive. Paradoxically, it was articulated through figures of racial degeneracy and corporeal mutilation that reflected specific anxieties: fears of famine, depopulation, political and economic impotence, and territorial dispossession can be detected between the lines of bombastic prose in praise of Italy's post-Unification population.

Biopolitical and Postcolonial Trajectories of Modernity

Early colonialism, aviation, mass political spectacle, public hygiene, science fiction, and cinema—these are the loci of Italian modernity that *Vital Subjects* traverses. If these are some of the fields in which a kind of self-conscious Italian modernity was articulated in this period, modernity is also inscribed in the two theoretical pillars that anchor the present study: the postcolonial and the biopolitical. In the first case, a postcolonial critique necessarily relocates Italian colonialism with regard to modern Italian nation building, shifting it from a marginal or epiphenomenal position to a more central, indeed constitutive, one. For Mellino, it is impossible to understand what was often cast as a triumphant Italian ascent toward capitalist modernity from the post-Unification period on without taking into consideration "[Italy's] underlying coloniality [and] its intrinsic racialization processes and racialized systems of domination" (87). This book therefore opens with a chapter

11 Italian irredentism (*irredentismo*) is a literary and political movement that emerged in the late eighteenth century and promoted Italy's wresting of the "unredeemed territories" from Napoleonic and Austro-Hungarian rule, in Piedmont, Liguria, and Tuscany and Trentino, Trieste, Istria, and Dalmatia, respectively. Irredentism is widely accepted as having prompted Italy's entrance into World War I. Its most famous post-World-War-I champion was Gabriele D'Annunzio. See Chapter Three of the present study.

on making Italians *vital subjects* through colonial technologies of power-knowledge, and concludes with a chapter on a film produced in celebration of liberal Italy's last attempt at colonial conquest, the invasion of Tripolitania and Cyrenaica (Libya) that began in October 1911. Indeed, part of what lends Italian racial discourse its specificity vis-à-vis other European traditions is the temporal and rhetorical proximity of its (belated) "birth" as a modern capitalist nation-state, the apex of positivist and formulations of biological race most memorably articulated by Alfredo Niceforo, Giuseppe Sergi, and Cesare Lombroso, and the height of Europe's so-called scramble for Africa.[12]

In the second case, biopolitics marks the transition from earlier forms of political subjectification to the making of the modern, or postliberal, political subject. For Suzanne Stewart-Steinberg, the figure of Collodi's Pinocchio is emblematic of the fact that post-Unification Italy was the locus of a modern rethinking of the national subject (and one that anticipated later theories of ideology by influential theorists such as Louis Althusser, Judith Butler, and Slavoj Žižek) insofar as the discourse around "making Italians" was profoundly aware of the performativity or fictionality of political subjectification (5–6). If biopolitics gives us a way of thinking about the modernity of the Italian political subject in this period, the goal of the following pages is to illustrate how it also offers us a productive hermeneutic alternative to the prevailing paradigm of anti-racism that has characterized most understandings of racial discourse in Italy since World War II.

The turn-of-the-twentieth-century Italian racial discourse that I explore in the coming pages is overwhelmingly preoccupied with the (re)productivity of life. Biopolitics names this a constitutively modern moment, when protecting and enhancing the life of a national population becomes the primary object of governmental calculation and political action. While the term biopolitics was coined in 1919 by Rudolf Kjellén, it wasn't until the mid-1970s when Michel Foucault would shed new light on it by tracing its emergence in the eighteenth century (and thus prior to Kjellén's naming of it) through what Foucault figured as the apex of biopolitical thought

12 Caterina Miele makes a similar case in her call for an archeology of racist discourse in Italy that takes into account the cultural specificity of Italy. Miele describes Italy's specificity with regard to racial discourse by identifying the importance of three critical features of Italian racial discourse: the contiguity of emigration and colonialism; the simultaneity of national unification with the height of positivist and biologist notions of race; and anti-southern racism as the necessary counterpoint to discourses of Italian modernity.

in Nazi Germany.[13] First and foremost in Foucault's seminal theorization of biopolitics was rejecting "life" as a taken-for-granted, natural category or presupposition and insisting instead upon its coming into being as a modern "problem," as a political question in itself that deserved consideration. Foucault understood biopolitics as a distinctively modern break with earlier forms of sovereign power through the emergence of what he termed "governmentality," or "the conduct of conduct." As Foucault explains, "Power is less a confrontation between two adversaries [...] than a question of 'government' [that] did not refer only to political structures or to the management of states; rather it designated the way in which the conduct of individuals or groups might be directed."[14] Whereas sovereign power followed a "deductive" logic in its effort to protect and maintain the juridical existence of the sovereign ruler, biopolitics and governmentality constitute a rethinking of modern power in terms of the productivity of the population (Foucault, *The History of Sexuality* 136). If sovereign power rested on the ability of the sovereign to *take* the lives of its subjects, governmentality and biopolitics are characterized instead by *guiding* and *augmenting* the (re) productive lives of subjects.[15]

Foucault's insights proved enduring and provocative, as they have been taken up over the past several decades by notable Italian political philosophers such as Giorgio Agamben, Roberto Esposito, Antonio Negri (writing with Michael Hardt), and Paolo Virno.[16] What Foucault highlighted, and

13 Michel Foucault's theorizations of biopower and biopolitics occur across a number of his published works and lectures from the mid-1970s. See, in this order: Foucault, *The History of Sexuality Volume 1* 142–143; Foucault, *Society Must Be Defended*; Foucault, *Security, Territory, Population*; Foucault, *The Birth of Biopolitics*.

14 Foucault, "The Subject and Power," in Rabinow and Rose, 126–144 and Foucault, "Governmentality," in Rabinow and Rose 229–245. The "Governmentality" lecture was delivered during Foucault's 1978 course at the Collège de France, which was published in English as *Security, Territory, Population*. In the midst of one of these lectures, Foucault remarked: "[I]f I had wanted to give the lectures I am giving this year a more exact title, I certainly would not have chosen 'security, territory, population.' What I would really like to undertake is something that I would call a history of 'governmentality'" (Foucault, *Security, Territory, Population* 108).

15 As Mitchell Dean explains, "The emergence of a form of rule that is distinct from sovereignty [governmentality] is also a rule over things but one that seeks to foster them, to increase the means of subsistence, to augment the wealth, strength and greatness of the state, to increase the happiness and prosperity of its inhabitants, and to multiply their numbers" (105).

16 Agamben, *State of Exception*; *Means without End*; *Homo Sacer*; Hardt and Negri,

what these subsequent theorists of biopolitics have illustrated in a number of re-readings ranging from revolutionary to tragic, is that, in order to defend the life of some, others must be deemed threatening and must therefore die. Indeed, in distinguishing this new form of power from the earlier sovereign right to "Take life and let live," Foucault famously distilled biopolitics to the edict, "Make live and let die." Most relevant for the present study is Roberto Esposito's theorization of the immunitary paradigm, which he has developed over the course of three works, *Communitas* (1998), *Immunitas* (2002), and *Bíos* (2004). Esposito identifies an immunitary logic at the heart of modern political categories, which accounts for biopolitics as paradoxically both "the protection and the negation of life," or, alternatively, the paradox that, "protection is the negation of life" (Campbell, "Interview with Roberto Esposito" 50). Esposito draws from biomedical and juridical language in order to tease out how immunity, as biomedical protection from contagion and juridical exemption from law has, over time, come to extend to "all those other sectors and languages of our life, until it becomes the coagulating point, both real and symbolic, of [all] contemporary experience" (51). Tracing an immunitary paradigm through its earliest expressions in the Hobbesian imperative of *conservatio vitae,* through the Hegelian dialectic and its fullest elaboration in Nietzsche, Esposito illustrates how immunity both undergirds and compromises community. Immunity is what guarantees the protection of a community, yet that protection, when carried to a certain point, risks perilously insulating life from both individual and collective existence: "what safeguards the individual and political body is also what impedes its development, and beyond a certain point risks destroying it" (51). As Timothy Campbell explains, Esposito argues that "the modern subject who enjoys civil and political rights is itself an attempt to attain immunity from the contagion of the possibility of community. Such an attempt to immunize the individual from what is common ends up putting at risk the community as immunity turns upon itself and its constituent element" (Esposito, *Bíos. Biopolitics and Philosophy* xi). This, for Esposito, is the constitutive moment of modernity. Esposito thus advances Foucault's reflections on the passage from sovereignty to governmentality as the linchpin of modernity by suggesting that both are steeped in an immunitary logic that necessitates both self-preservation and self-negation (xii).

Commonwealth; *Multitude*; *Empire*; Virno, *A Grammar of the Multitude*. See also: Berardi; Bazzicalupo, *Biopolitica*; "The Ambivalences of Biopolitics"; Forti; and Lazzarato, *Lavoro immateriale*.

Labor and Biological (Re)productivity

A few notes on Marx's seminal analysis of labor and (re)production will bolster this discussion of Foucauldian biopolitics and Esposito's immunitary reading of it, and offer some crucial theoretical distinctions toward an analysis of how biopolitical rhetorics of (re)productivity shaped Italian racial discourse in the post-Unification period. The "making" of Italians as *vital subjects* by Franchetti, Mantegazza, D'Annunzio and Pastrone relies upon rhetorics of (agricultural) labor productivity and biological reproduction that sometimes overlap; these terms pertain unequivocally to the terrain of biopolitics, given that the (re) productivity of bodies conceived of as elements of a population is one of its constituent concerns. Though Marx posited labor as the basis of human and societal existence, he famously dedicated little space to the material processes of biological reproduction that create the conditions for such existence and for the system of capitalist production upon which it depends. Marx's shortcomings in this regard were themselves subsequently "reproductive," as they sparked an influential wave of Marxist feminist thought, which aimed to redress such critical lacunae in Marx's analysis of capital.[17] In his genealogy of late twentieth-century biopolitics in Italy, Andrea Righi positions the theoretical work of the post-1968 neo-feminist organization *Lotta Femminista*, or the Wages for Housework Movement, as a critical intervention, as it sought to include the non-waged labor of housewives and question the concept of female emancipation through work (58). Furthermore, *Lotta Femminista* repositioned human or biological reproduction vis-à-vis capitalist production, inverting the terms that made the former a mere auxiliary of the latter. Each of the terms of interest here (labor and [re]production) contains a complex dual and even tripartite structure; as such, Marx's dialectical argumentation requires some patient elucidation.

Marx memorably begins *Capital* with the commodity, "an external object, a thing which through its qualities satisfies human needs of whatever kind" (125). The commodity, from boot polish and buttons to silk and gold, is a product of human labor and a material manifestation of use-value, exchange-value, and value. The dialectical relationship between a commodity's usefulness to oneself and others, its value as a unit of exchange, and the labor time necessary to produce it leads, following Marx's quasi-Hegelian mode of argumentation, to a duality in his definition of labor. If value is socially

17 For a survey of these arguments, see: Hearn; Himmelweit.

necessary labor time (or the "[average] labor time necessary to produce [commodities] under given conditions of labor productivity," as Harvey (25) explains), Marx identifies two kinds of labor embodied in the commodity as a unit of exchange: concrete labor (heterogeneous, from mining to weaving, etc.) and abstract labor (homogeneous, all products of labor considered in general). Like the commodity, which embodies multiple types of value, labor contains both concrete and abstract forms (the latter is itself synonymous with value, as socially necessary labor time). This discussion of the social relations of labor concealed by the "mysterious" commodity form will bring Marx to his groundbreaking analysis of commodity fetishism, which will become pertinent later on in the discussion. For now, one more note on the relationship between the commodity and the labor required to produce it. To discuss the way in which the capitalist searches for a commodity whose use-value is itself a source of value, Marx introduces another, third term: *labor-power* (270). Labor-power, defined as the aggregate physical, mental, and human capacity for labor, is thus exchanged as a commodity, as the sole commodity that the laborer owns (this is naturally the case only where laborers are 'free,' rather than enslaved). The laborer may sell this commodity to the capitalist, who in turn becomes the owner not of the laborer himself (as in slavery), but of his capacity to produce. But the laborer, a living being, must himself be sustained, indeed reproduced, in order for the value of labor-power to remain constant. "[A] definite quantity of human muscle, nerve, brain etc. is expended, and these things have to be replaced" (274). Enter again the value of all of those other commodities that replace these expended energies of the laboring body: food, clothing, fuel, housing, and so on (275). In this discussion of labor-power, then, one notes the passage from the realm of the capitalist production of commodities (including labor-power qua commodity) to the realm of the nourishment and reproduction of laboring bodies. A curious commodity, precisely because it is "a living agent of fermentation," labor-power thus names a site where the economic and physiological coalesce (292). Furthermore, while Marx discusses labor-power as an abstraction, as an aggregate of the human capacity for labor, it is never fully removed from the concrete labor and the individual body of the worker, as in Crispi's formulation with which this chapter began, which hails both the individual emigrant body and the collective "fruits of [Italian emigrant] labor." Indeed, for Righi, labor-power is precisely the "crucial theoretical point," which "opens the gates to the biopolitical dimension." (57)

Across Marx's works but particularly in *Capital*, reproduction refers most often to the ability of social systems (such as capitalism) to reproduce themselves, to maintain their existence through processes that define and determine them (Himmelweit 197). Such *social* reproduction is thus tied to Marx's materialist conception of history, according to which people, not nature, are the agents of history. Yet his materialism sometimes falters, particularly when it comes to fully considering the capitalist organization of human biological reproduction. As Susan Himmelweit alleges, "[Marx's] failure to analyze the social relations of reproduction leaves incompletely fulfilled [his] own aim of offering a materialist account of the capitalist mode of production" (210). On the few occasions when Marx discusses it head-on, human reproduction is left to a kind of naturalized or psychological impulse for procreation on the part of the worker, which the capitalist only need harness. In *Capital,* he writes, "The maintenance and reproduction of the working class remains a necessary condition for the reproduction of capital. But the capitalist may safely leave this to the worker's drives for self-preservation and propagation" (718). This tension between "natural" and social reproduction had appeared earlier, in Marx and Engel's *The German Ideology* (1845–1846), where the authors stated that "the production of life, both of one's own in labor and of fresh life in procreation, now appears as a twofold relation: on the one hand as natural, on the other as a social relation" (43). In *Capital,* Marx's consideration of the social relations of production ambiguously brings together, without fully resolving, the biological reproduction inherent in labor-power and the social reproduction of capital and capitalist relations themselves. For instance, Marx writes:

> The conditions of production are at the same time those of reproduction. No society can go on producing, in other words no society can reproduce, unless it constantly reconverts a part of its products into means of production, or elements of fresh production. All other circumstances remaining the same, the society can reproduce or maintain its wealth on the existing scale only by replacing the means of production which have been used up [...] by an equal quantity of new articles. (711)

Here, one appears to be in the realm of social reproduction, but one can also see how this realm is never entirely separable from what Marx figures as "natural" human reproduction. Indeed, in the pages that follow this passage, he refers to several additional meanings of reproduction: from the biological reproduction of the working class (noted above) to the reproduction of capital by the worker; reproduction as "a mere repetition of production," or "simple reproduction" (712; 715); the maintenance of labor-power according to which "the muscles, nerves, bones, and brains of existing workers [are reproduced]" by the capitalist (717); and so on. These varied definitions, argues Jeff Hearn, are ultimately subsumed by the social reproduction of capitalism as the primary form of reproduction. Indeed, this was precisely where *Lotta Femminista* intervened when they inverted Marx's logic, suggesting instead that the work of reproduction is carried out not by male workers, but instead mostly by women (who are virtually absent from Marx's discussion of reproduction, except presumably as passive recipients of "the worker's drives for self-preservation and propagation"). As Righi puts it, "Domestic work [...] is what makes labor possible. [...] As its prerequisite, reproduction is the condition of possibility for the capitalist realization of profit" (58).

What does Marx's analysis reveal with regard to the biopolitical dimension of Italian racial discourse and the making of *vital subjects* with which this book is concerned? The social historian, following Italy's preeminent Marxist philosopher Antonio Gramsci, might hasten to clarify that the largely agricultural economy of post-Unification Italy was far from the highly industrialized British context that Marx so attentively described. Nevertheless, labor-power provides a clear theoretical frame in which social and economic relations converge with biological concerns about the maintenance and reproduction of human life. Again, this is precisely the realm of biopolitics: when the primary rationale of government becomes managing the circulation of individuals, conceived of as elements of an aggregate national population. Indeed, it is in the field of political economy that Foucault argued life was introduced into history, thus becoming a political problem rather than a "natural" given. As soon as economy (the government of the family) and politics (the government of the polis) became imbricated with one another, Foucault argues that new techniques of power emerged. The rationality informing these techniques was governmentality, and addressed "the correct manner of managing individuals, goods and wealth within the family (which a good father is expected to do in relation to his wife, children and servants), and of making the family fortunes prosper – how to introduce this meticulous attention of the father towards his family into the management of the state?" (Foucault, *Security, Territory, Population* 94–95). Modern political economy, Foucault contends, grew out of the Latin *oeconomia*, the management of the *oikos*, or home. Indeed, in Italian, as in English, the term *governare* contains this split or dual meaning (both semantic and gendered): *governante* refers to both statesman or ruler *and* domestic governess or housekeeper. Foucault's governmentality, as a biopolitical apparatus, is, according to Maurizio Lazzarato, a "political economy of forces" ("From Biopower to

Biopolitics") that is both quite similar to and yet ultimately different from Marx's analysis of labor-power as living labor (as opposed to the dead labor of capital, which "vampire-like, lives only by sucking living labour, and lives the more, the more labour it sucks" (Marx 342), in the first case because both Marx and Foucault are concerned with modes of coordinating the economic and ontological relationships between living men and things, with the aim of extracting surplus value, or power. Yet Lazzarato also points out that, in the second case, Foucault accused Marx of reducing the entire, dispersed field of power relations to binary relations between capital and labor, making these relations "the source of all social dynamics" (11). Indeed, Foucault would point out how Marx's model of antagonistic class struggle may itself get caught up in the racist drift of biopolitics, insofar as the basis for such a model is always already informed by the older logic of race war.[18] Rather than a war of position between two opposing camps, Foucault's governmentality is instead a diffuse material field in which natural resources, commodities, circulation and commerce are engaged, but also where urban planning, public health, and the perceived "fitness" of the population come into play.

Biopolitical Rhetoric and "The Literariness of Life"[19]

A postcolonial, biopolitical approach—rather than a strictly anti-racist one—reveals how race and colonialism were more central to Italian nationalist rhetoric prior to the fascist politics of demography and empire than was, until the last decade or so, acknowledged in conventional scholarly accounts.[20]

18 As Warren Montag writes, "For Foucault, it is not just that totalizing theories of opposing camps [classes] locked in a war of position are inaccurate descriptions of the real complexity of the field of struggle, it is even more that they tend to replicate the form, if not the content, of the older notion of race war" (116–117).

19 See Hunt and Rudolf. The article lacks page numbers; henceforth I cite instead from its numbered paragraphs.

20 Notable exceptions are: Silvio Lanaro, and Alberto Banti's study, which explores how the myth of the nation gathered new force around the Risorgimento as its creators drew from a range of existing notions, from the Christian tradition (sacrifice, martyrdom, purity) to kinship relations (mother-nation, community of brothers, etc.). Banti explains how beliefs in a nation constituted by consanguinity, or blood relations, and rooted in a circumscribed territory (from town to region and eventually to nation) circulated in Italy from the end of the eighteenth century. Banti's thorough study of the Italian nation as a rhetorical construction seeks to explain how a handful of intellectuals gave shape to an idea that men (and, though quite parenthetically, women) risked their lives defending.

Furthermore, the broad field of contemporary Italian political thought about biopolitics that Foucault inspired has yet to thoroughly account for the role that race thinking, colonialism, as well as the related experience of mass emigration played in the historical trajectory of biopolitical thought and practice in Italy.[21] One aim of the close readings that follow is to link this influential current of biopolitical thought by contemporary Italian political philosophers to the post-Unification convergence of three "problematics of the population": the southern question; early, pre-fascist colonialism; and mass emigration, which lent lasting shape to Italy's particular brand of racial discourse. While one might name several other significant thresholds in the development of current biopolitical thought and practice in Italy, focusing attention on the years around Italian Unification affords access to a critical prehistory of some contemporary phenomena that have received attention in critical thought about biopolitics, from Hardt and Negri to Esposito: regimes of the capitalist exploitation of human labor-power and rhetorics of

He describes in great detail the evolution and reception of the nationalist canon, though his treatment of the racialist components of Risorgimento nationalism serves primarily to illustrate his thesis about the symbolic potency of kinship (Banti, *La nazione del Risorgimento* 156–165).

21 For a reading of Giorgio Agamben's *Homo Sacer* with regard to Italian colonialism in Libya, see Atkinson. Ann Stoler was one of the first readers (or listener, as the lectures were yet unpublished in English or French and Stoler came across them for the first time as "scratchy cassette recordings") of Foucault's 1975–76 *Society Must Be Defended* lectures to illuminate how they contributed to a colonial reading of Foucault. In *Race and the Education of Desire*, which appeared in 1995 (nearly ten years before the English publication of *Society Must Be Defended*), Stoler provided for English readers the most thorough reading to date of Foucault's lectures on how biopower inscribed racism in the modern state. Her project aimed to resituate Europe's colonial history within the frame Foucault had sketched of the biopolitical nation-state. Unsatisfied with how race and colonialism figured in Foucault's analyses, but inspired by the potential they presented, Stoler made the case for placing race and other colonial regimes of power/knowledge at the center of bourgeois sexuality and statecraft. It is of note that, as the entirety of the lectures had not yet been transcribed into French (and, as we've seen, their complete English translation was not to appear until 2003), Stoler relied upon the text of an unauthorized Italian translation of the lectures that appeared in 1990 as *Difendere la società* and was promptly removed from bookshops at the request of the Foucault estate (Stoler, *Race and the Education of Desire* 57). Stoler is credited with having been the first to publish in English on the lectures, and her quotations of the text are derived from an unattributed translation of the Italian text. The introduction of Foucault's most sustained analysis of biopolitics to an English-reading public, then, was itself effected through a linguistic detour through Italy. See also Scott.

immigration as biological contagion, to name just two.[22] In addressing why Foucauldian biopolitics has received so much attention from Italian thinkers over the past few decades, Roberto Esposito asserts:

> It's true that Italy, perhaps more than any country, is the place in which Foucault's reflections on biopolitics [...] have been extended with more breadth and originality [...]. Why? [...] Italy is a country on the frontier, not only in a geographic sense, but also culturally, between different worlds, between Europe and the Mediterranean, between North and South [...]. Italy is traversed but also in a certain sense constituted by this fracture. Perhaps the sensibility to a theme such as biopolitics may be linked to this liminal condition of the border, for biopolitics is also situated at the intersection between apparently different languages such as those of politics and life, of law and of anthropology. (Campbell, "Interview with Roberto Esposito" 49)

If Italy has been a particularly fertile site for recent theorizations of biopolitics, it is at least in part because Italy's history as a (post)colonial nation-state has been profoundly shaped by the constitutive fractures that characterized Italy during the crucial years between its political formation as a liberal nation-state and the rise of fascism. An even more detailed snapshot of such fractures than the one offered by Esposito would include the linguistic pluralism of Italy's diverse regions as well as other, similarly well-traveled scholarly itineraries of Italian disunity. The coming chapters illustrate how such fractures are 'mended' rhetorically (and somewhat paradoxically) by a brand of colonialism aimed at restoring missing "arms" (as in Crispi's formulation) to a mutilated geopolitical body; formulations of emigration as responsible for carrying the "arms" of millions of productive worker-citizens to far away nation-states; and decrying the threat to national unity presented by the entrenched divide between Italy's northern and southern regions with regard to capitalist industrialization and the potential for its exploitation by a newly centralized state.

Volumes have been dedicated to examining the contemporary effects of biopolitics in a number of national and intellectual contexts ranging from ecology and bioethics to political science and philosophy, particularly over the

22 For a twentieth-century genealogy of biopolitics through the work of Antonio Gramsci, the workerist feminism of *Lotta femminista*, and Pier Paolo Pasolini, see Righi.

past few decades.[23] In the present study of post-Unification Italy, biopolitics is taken up in two overlapping ways. Historically, biopolitics names a distinctively modern mode of power, or governmentality, aimed at enhancing and protecting the life of the population. Biopolitics as a mode of power emerged in the eighteenth century and continues, some would argue with increasing intensity, to the present day, with Nazi thanatopolitics representing one of its most grisly, and yet to be entirely overturned, expressions.[24] Foucault coined governmentality in order to explain how the ways in which human beings are made into political subjects are irreducible to the workings of the state, but are instead carried out by a range of actors and institutions that ensure the "conduct of conduct." Public hygiene, charitable institutions, demographic statistics, insurance, individual and collective savings, and other safety measures aimed at reducing the aleatory aspect of the life of the population are some of the technologies of governmentality that characterized the emergence of the biopolitical era in which we continue to live.

If biopolitics refers to an historically specific mode of power, my reference to a biopolitical approach to post-Unification cultural production signals a focus on questions of language and representation. Indeed, the biopolitical technologies listed above (demography, public hygiene, etc.) rely above all upon strategies and modes of representation. As Alastair Hunt and Matthias Rudolf argue:

> [L]ife, both in the work of Foucault and Agamben as well as other theorists of biopolitics, becomes accessible to biopolitical intervention, not as such, but through its entry into language and representation. [...] [P]olitics intervenes on life through the production, regulation, and manipulation of figural regimes—the statistics, estimates, data, totals, and sums that represent the life of the population at a general level [...]. What is decisive is not any positive knowledge of the relation of power and life, but rather the mere facticity of the substitution of language for life that enables that intervention in the first place. (20)

Hunt and Rudolf explore how, in the decades since its explosion after Foucault and Agamben, biopolitics has been used most productively to formulate a number of important questions within the social sciences, as well as history

23 For an excellent survey of the broad range of definitions of biopolitics see Lemke.
24 For the argument that we have yet to entirely overturn some of the fundamental presuppositions of Nazism, see Esposito, "Nazism and Us."

and political philosophy. They demonstrate how biopolitical theory can benefit from a deeper engagement with literary culture, pointing out that "even the most assured representation of life turns upon an irreducible 'literariness'" (4). Accordingly, the focus of this book is on discursive mechanisms and "figural regimes" that produce or destroy human life in order to figure Italians as modern, racialized political subjects. At stake for the post-Unification thinkers considered in this book's upcoming chapters was how to make the biological lives of Italians, as members of a newly unified national population, (re)productive. As Nicole Shukin demonstrates in a different but related context, a biopolitical approach to post-Unification cultural production means that "textual logics of reproduction can no longer be treated in isolation from economic logics of (capitalist) reproduction" (20). The figurative "fruits" of Italian labor, which early colonial proponents such as Crispi imagined would be born from demographic colonization, and his recourse to the metaphor of the body politic and the "organic" state to describe Italy's (bio)political economies of emigration and colonization attest to the centrality of a certain literariness inhering in racial and biopolitical discourse.

Situating racial discourse in the historical and theoretical contexts of biopolitics reveals a number of meanings of race in post-Unification Italy that might have been left unexamined with recourse to a strictly anti-racist analytic that instead focuses attention solely on instances of stereotypes or explicit racialism. While analyzing the construction of difference or otherness would undoubtedly be productive in the case of explicit attempts to represent racial and/or colonial otherness—the pages of weekly periodicals such as *Illustrazione italiana* and *Domenica del Corriere*; the "southernist" literature of Giovanni Verga, Leonardo Sciascia, and Giuseppe Tomasi di Lampedusa; the influential craniometry and criminology of Cesare Lombroso; the overt racism of Futurist poetics; and lesser-known fields such as early colonial photography all come to mind as examples—this book focuses on the constitution of racialized Italians in the context of biopolitics.[25] Many Italian texts

25 On the relationship between the commodification of the primitive in southernist literature, theater, and early film culture, see Bertellini, *Italy in Early American Cinema* 81–92. See also: Moe, "The Geographical Poetics of Giovanni Verga"; *The View from Vesuvius* 250–289; Rosengarten. One "accidental enthnographer," Rosalia Pianavia Vivaldi, lived in Eritrea from 1893 to 1896 and later published her colonial memoir *Tre anni in Eritrea*, which blends ethnographic description, autobiography, and a number of photographs of Eritreans and Italian colonials taken by Pianavia Vivaldi herself. See: Sòrgoni, "Italian Anthropology and the Africans." Photographers Luigi Naretti and Giovanni and Francesco Nicotra "captured" arrested rebels accused of aiding in the

written in this period produce physiological expressions of race, focusing on the making of Italians through invisible elements such as blood, organs, and vital fitness. Often, more important than skin color or physiognomy is a unique preoccupation with which bodies are reproducing, laboring or sacrificing themselves in battle, and where—be it in the Americas, Italy's southern regions and islands, the new Eritrean colony in the Horn of Africa, the mutilated territories of the Adriatic, or the "lost" ones of the Mediterranean (Tunisia and Libya). Given the breadth of cultural expressions of racial representation in post-Unification Italy, Ann Stoler's observation is apposite:

> Racial discourse is neither always a tool of the state nor always mobilized against it. Racial discourses diffuse over a broad field. Their genealogical histories should track their 'spaces of dissention' and unique sites of dispersion. [...] Racial discourse [...] accrues its force not because it is a scientifically validated discourse but just the opposite. It is saturated with sentimentalisms that increase its appeal. (*Carnal Knowledge and Imperial Power* 159)

These "unique sites of dispersion" include not only racial science, but also domains such as literature, politics, popular hygiene, and cinema. Rather than analyzing the many tropes of racial otherness that pervaded post-Unification Italian culture, the present study asks: what are the biopolitical coordinates of the racialization of Italians before fascism, and how have these boundaries helped stake out the parameters of modern Italian nationalist discourse? Taking into account Hunt and Rudolf's claims about biopolitics and the "afterlives of [literary] romanticism" and Stoler's claim about the "sentimentalism" of racial discourse, what sorts of romantic sentimentalism does Italian racial discourse attempt to harness and how are such appeals to affect grounded in the biopolitical? The answer to such questions lies in the specificity of an *italianità* (Italianness) that was defined in terms of what Mark Choate has aptly called a "sentimental tradition," rather than in juridically binding formulations of citizenship, for example. If, following Stoler, racial discourse harnesses affect for its strength, and if, as a result of Italy's heightened concerns with emigration and colonialism in this period,

defeat of Italian troops at Dogali in 1887 for the Treves' illustrated weekly *Illustrazione italiana*. For more on early colonial photography, see: Forgacs; Palma, "The Seen, The Unseen, The Invented"; *L'Italia coloniale*. For a gendered analysis of the influence of Lombroso on Italian sexual politics, see Gibson.

Italianness was forged above all through sentiment, one wonders about the inextricability of one term (race) from the other (*italianità*). In fact, the book that is widely recognized to have been, alongside Carlo Collodi's *Pinocchio*, among the most significant examples of literary nation-building in post-Risorgimento Italy, and that was responsible for the sentimental and patriotic education of generations of Italians, is Edmondo De Amicis' *Cuore* of 1886. Historian Giuseppe Maria Finaldi has argued that *Cuore*, as "almost a checklist [...] describing Italy's ideal citizen," should be connected to a budding post-Unification "culture of colonialism" for the ways in which it foresees not only a population bound by the blood of genealogy and shared sacrifice residing within particular territorial limits, but also the expanding of those borders through the shedding of new blood, which would allow Italy to, in De Amicis' words, "live and expand, calm in the majesty of [her] right and strength" (Finaldi 48–49). Moreover, the fact that at the center of *Cuore* is a relatively lengthy short story titled "From the Apennines to the Andes," which foregrounds Italian patriotism and sacrifice in a tale about emigration to the Americas, adds yet another layer to *Cuore*'s relation to the "culture of colonialism" described by Finaldi.[26] Nineteenth-century emigration to the Americas was inextricably linked to the liberal-era colonial schemes in both the Horn of Africa—where Francesco Crispi, with the aid of Franchetti, initially supported "demographic" agricultural settlements of Italians—and in South America, where Luigi Einaudi sought to harness "spontaneous" Italian emigrant colonies as an alternative to what he described as the "insanity" of African colonization à la Crispi (Choate, "From Territorial to Ethnographic Colonies" 68–70).

In the period following Unification, the dreams of the nationalizing project of the Risorgimento began to disintegrate, provoking the elaboration of ever-new imaginative strategies for the production and management of the Italian population. A focus on this period, when Italy was most self-consciously if somewhat desperately clamoring to join the triumphant European march toward capitalist modernity, enables an examination of how the very ideological structures of the modern nation-state—before

26 The thirty-page "From the Apennines to the Andes" tells the story of Marco, a thirteen-year-old Genovese son of a manual laborer whose mother had emigrated two years earlier to Buenos Aires to seek well-paid domestic work. Marco embarks alone on a quest for his lost mother across the Atlantic, where he is generously aided by members of the "spontaneous colonies" of emigrant Italians in Argentina (De Amicis, "Dagli Apennini alle Ande").

their "infection" by fascism—were erected and reinforced on the backs of an explicitly biological population. In less than forty years, between 1880 and 1915, statisticians estimate that 13 million Italians left home, distinguishing Italy as the nation-state to undergo the largest emigration in world history (Choate, *Emigrant Nation* 1). Indeed, under Prime Minister Crispi (1887–1891; 1893–1896), Italy's largest export was neither food nor fashion—it was *manodopera,* human labor-power. The three biopolitical fields, the southern question, migration, and colonialism, in which Italian racial discourse took shape, converged most markedly during the post-Unification period, when Italy experienced the mass exodus of peasants in search of work, land, and life elsewhere. Significantly, this mass migration was termed in official and popular discourse as either "hemorrhage" or "bloodletting," thus bringing into focus the centrality of the relationship between the liquidity (in terms of fluids like blood and sperm) that courses through Italian bodies and the oceans and seas that conduct those bodies elsewhere, and territory: the "lost" (read formerly Roman) lands of Libya, or the "unredeemed" territories of the Adriatic were at risk of falling into the wrong hands, thereby further "mutilating" an already tentative disposition of bodies and territory.[27]

In addition to shifting the focus on structures of racialization from fascist to liberal Italy, in the chapters that follow, I shift the focus from explicitly racialized objects to the making of Italians as racialized subjects. As Cristina Lombardi-Diop has argued:

> On the whole, today's social critique of racism [in Italy] has dedicated little attention to the ways in which race, as a system of differentiation, "shapes those on whom it bestows privilege as well as those it oppresses." Moreover, studies on the construction of modern Italian identity have paid little attention to the impact of racial self-definitions and self-perceptions. Scholars, with few exceptions, have not interrogated the racial assumptions that have structured and supported the idea of Italianness as racially coded. ("Postracial/Postcolonial Italy" 176)[28]

27 Uli Linke ties recent anti-immigrant rhetoric in Germany to firmly embedded mythologies of blood through the former's metaphorical recourse to liquidity: "flood," "stream," "wave," "flow," etc. She suggests that "metaphors of water may well be circumlocutions for blood, [which] [...] finds expression in terms of other fluids or liquid substances: foam, sweat, whirlpool, river, stream" (*Blood and Nation* xi).

28 Lombardi-Diop cites Ruth Frankenberg's article "White Women, Race Matters." *Theories of Race and Racism: A Reader.* Ed. Les Back and John Solomos. (London and New York: Routledge, 2006) 519–533.

In order to contextualize the emergence of more violently racist language—particularly vivid examples include the language of high fascism during the invasion of Ethiopia in 1935–1936 and the racial laws of 1938–1939, or the virulently anti-immigrant vocabulary that has been most audible in the discourses of the extreme right since the 1980s—this book interrogates the discursive constitution of the subjects from whom such language eventually flows.[29] Simply put, Italians are interpellated by the very logics of racialization that they employ. This presented some difficulties for many Italian thinkers in the post-Unification period, as within their nationalist narratives in praise of the Italian people was embedded an acute awareness not only of Italy's marginal position with regard to stronger European powers, but also of the related risks of racial degeneracy and explicitly racist discrimination against Italians in the Americas.[30] Racialization thus blurs a distinction between subject and object, as the subject who racializes must him/herself be racialized. Such a contention is by no means an attempt to obscure the importance of racial discourse to the interplay of power-knowledge: indeed, in the case of race, what sets "subject" apart from "object" are power relations.[31] Still, the Italian case is an interesting one precisely because of the threat of Italian racial inferiority that was inscribed in both Risorgimento-era and post-Unification texts by Italy's subordinate position vis-à-vis established western European powers and the racialization of Italian emigrants in the Americas.[32] As Stewart-Steinberg has argued in a somewhat different but related context:

> Anxiety does in fact describe the post-1860 moment. [...] This anxiety [...] is fundamental to Italian modernity rather than an impediment to it. [...] [T]he formulation of an Italian national self was predicated on a

29 Though Italy's first anti-Semitic laws were passed in 1938, laws prohibiting sexual relations between Italian citizens and colonial subjects were in fact passed in the African colonies a year earlier, in 1937. See: Barrera, "Sex, Citizenship, and the State" 157.

30 In 1891, eleven Italians were lynched in New Orleans, and roughly thirty were killed by a mob in Aigues-Mortes, France. Mark Choate writes: "This drew uncomfortable parallels with the persecutions of African, Chinese, and South Asian minorities worldwide" (*Emigrant Nation* 15). News of these deaths traveled back to Italy, adding to the racial anxiety that characterized post-Unification emigration debates.

31 Philomena Essed and David Theo Goldberg argue, "All variations on and through 'race' serve as codes and manifestations of power more generally, and they so often factor more or less quietly or more or less explicitly into a complex of causes for political, economic and social conditions" (4).

32 For more on Italian "self-Othering" around the Risorgimento see Patriarca, "Indolence and Regeneration."

> language that posited marginalization and powerlessness as fundamental aspects of what it meant to be modern Italians. (2)

Perhaps the most paradigmatic example of the anxiety that characterized the national project was the southern question, arguably the leading question for policymakers after Unification. At stake was how to bring southern Italians—who were depicted in what became known as *meridionalista* ("southernist") literature as either hapless victims of history plagued by poverty, illiteracy, superstition, and other forms of "backwardness" or as criminally, culturally, and/or racially resistant to the rationalist progress of liberal-capitalist development—into the fold, how to represent southerners in a newly nationalized political community. This project was intimately connected to another perceived fragmentation: the loss of Italy's brothers and sons to transatlantic emigration. Panic about emigration resulted in an unprecedented campaign by the new national government aimed at conquering the "hearts and minds" of Italians from a distance.[33] While some thinkers devised strategies to instill *italianità* from across the Atlantic Ocean, others set their sights upon colonies across the Mediterranean to remedy this fracture. Race, as a way of thinking about the relations between peoples and lands whose primary mode is one of fragmentation, thus takes on a new shape when inscribed within such a rhetorical economy, as it cannot be explained as merely an ideology of the dominant, nor as innocuous, dated make-believe.

On the eve of the nineteenth century, Ugo Foscolo, who is considered, along with Giacomo Leopardi and Alessandro Manzoni, one of the foremost representatives of Italy's Risorgimento-era literary Romanticism, put the following invective into the mouth of his tragic patriot-hero Jacopo Ortis:

> I tuoi confini, o Italia, son questi! ma sono tutto dì sormontati d'ogni parte dalla pertinace avarizia delle nazioni. [...] E verrà forse un giorno che noi perdendo le sostanze e l'intelletto e la voce, sarem fatti simili agli schiavi domestici degli antichi, o trafficati come i miseri Negri (*Le ultime lettere di Jacopo Ortis* 132)

> [Oh Italy, your borders are these, but every day they are overcome by the persistent avarice of all nations. [...] Perhaps the day will come when we, having lost our possessions, our intellect, and our voice, will be made

33 See also: Choate, *Emigrant Nation*; Wong.

> similar to the domestic slaves of the ancient lords or traded like wretched negroes] ("The Last Letters of Jacopo Ortis" 92)

As this canonical text of early Italian nationalism illustrates, nearly a century before the southern question and consequent emigration laid bare how Italians were faced with explicit racial discrimination in the Americas, the breaching of their borders by French and Austrian occupiers drew uncomfortable parallels between Italians and African slaves.[34] Rhetorics of dispossession and loss (in terms of territory and labor-power, which, as we have seen, is both economic and physiological) that emerged most markedly in liberal-era discussions of emigration and colonialism were thus already inscribed in Risorgimento-era literary nationalism. Foscolo's narrative of Italian dispossession ("[your borders] are overcome by the persistent avarice of all nations") and racial degeneration ("having lost our possessions, our intellect, our voice, [we] will be [...] traded like wretched negroes") highlights the loss, rather than the fundamental absence, of Italian territorial and corporeal identity. An identity lost can be found again; an absent identity was never there in the first place.[35]

Race and Biopolitical Fantasy

Esposito's discussion of Italy's constitutive fractures as an explanation for why biopolitics has found such a wealth of theorizations there, and what I have been suggesting is a rhetorical mending of territorial and corporeal breaching, dismemberment, dispossession, and loss performed by biopolitical discourse in the post-Unification period raise the question of how biopolitics relates to psychoanalysis, and more specifically to the mechanisms of fetishism and ideological fantasy. Foucault, and consequently many of the theorists of biopolitics who followed in his wake, notoriously claimed that psychoanalysis, as a constitutive part of the larger "explosion" of the discourse of sexuality,

34 For more on the connections between African slaves and Italian immigrants, see Wong 113–148.

35 Dominick LaCapra calls for greater clarity in distinctions between absence, which is foundational and transhistorical, and loss, which is instead historical. He warns that a conflation of loss and absence, particularly surrounding such traumatic events as the Shoah and South African apartheid, can reproduce the conditions that cause historical traumas in the first place, thereby preventing the acting-out and working-through required to address them. See his essay: "Trauma, Absence, Loss."

was but one of several technologies to emerge in the passage from sovereign power (the power to make die, *faire mourir*) to biopolitics (as making live, *faire vivre*). With biopolitics, as "the people" become subjects of power, the object of power becomes "the population." This objectification requires apparatuses of governmentality (public hygiene, charitable institutions, demographic statistics, etc.) aimed at the management of life, among which psychoanalysis emerges as, in Anna Kornbluh's apt quip, a "regional manager, the storied lieutenant of 'a general deployment of sexuality'"(17). Psychoanalysis is thus but one mechanism of governmentality. Seeking to quell the rift between the biopolitical paradigm and psychoanalysis with regard to their distinct, if parallel, theories of the (political) subject, Eric Santner has called convincingly "to put Freud and Foucault on the same team" (xiii). Santner argues that, given the centrality of the body of the king as the *symbolic* basis of sovereignty to Foucault's analysis, the biopolitical paradigm has more in common with psychoanalytic theories of the subject than has been conventionally acknowledged. Santner, like Stewart-Steinberg, explores the ties that bind physical bodies to the symbolic order of politics. He argues that with "the people," newly vested with sovereignty in the wake of the French Revolution, is born an anxiety about the specter of the fragmentation of this biopolitical body. Anxiety also results from the mirror stage, as Jacques Lacan calls it, as the threat of bodily fragmentation ensues from one's encounter with a whole image reflected back to him or her. The consequences for both models of sovereignty theorized by Foucault are thus: the body of the king as the symbolic basis of sovereignty is in fact a mortal body, on the one hand, and, on the other, the displacement of sovereignty onto the popular body intensifies the threat of fragmentation by placing political power into the indistinct and contested hands of "the people." "The people" as the biopolitical agent of power is a constitutively fractured entity; its singular, unified name may belie a fundamental emptiness.

A figure simultaneously empty and fundamentally split, "the people" as the unified subject of sovereign power may be considered the ideological fantasy par excellence of all liberal democratic nation-states.[36] For cultural theorist Slavoj Žižek, ideological fantasy—the illusion which structures "our real, effective relationship to reality"—reconciles Marx's argument about how commodity fetishism disguises our social relation to labor through the exchange of things (summarized in the maxim "they do not know it, but they are doing it") with

36 For an analysis of the elusive notion of the people as the basis for American democracy, see Frank. For a distinction between "the People" as the basis of sovereignty and its necessary corollary, "the people" as bare life, see Agamben, *Means Without End*.

the psychoanalytic structure of fetishistic disavowal, famously formulated by Octave Mannoni as "Je sais bien, mais quand même" ("I know very well, but all the same") (18). Similar to the Freudian fetishist, who turns to alternative objects (the foot or the shoe) to stand in for a maternal lack (the phallus) and as such is able, through disavowal (*Verleugnung*), to maintain two contradictory beliefs (that woman both has and does not have a penis), Žižek argues that what Marx's fetishist overlooks or misrecognizes is not reality itself, but the illusion which is structuring his or her reality. Individuals under the spell of commodity fetishism need not believe in the illusion of the intrinsically magical nature of money—they know, claims Žižek, that a commodity such as money merely represents a social relation of exchange, but nevertheless they behave *as if* it were the embodiment of wealth. Žižek explains, "The fundamental level of ideology [...] is not of an illusion masking the real state of things but that of an (unconscious) fantasy structuring our social reality itself" (33). Ideology, far from false consciousness, binds subjects (and texts) to the necessary illusions (the body of the king; the sovereignty of "the people") that structure their reality.[37] Yet the illusion has concrete effects; so goes the saying, money makes the world go round. For Stewart-Steinberg, "the very idea that subjects are bound to the state in an ideological, that is, imaginary and yet material, manner coincides with the birth of another form of bondage, a disciplinary one founded on the production, management, and enhancement of life" (10). She goes on to discuss how biopolitics, as a means of "making live" through governmentality rather than a pre-modern notion of sovereignty exemplified by the king's sword (instead, a means of "making die"), does not entirely displace sovereign power, and asks instead about the interaction in post-Unification Italy between these two forms of power. She suggests that the project of "making Italians" constitutes an attempt to negotiate between these two modes. Indeed, Foucault's analysis of state racism in his canonical lectures on biopolitics at the Collège de France suggests that biopolitics does not rule out the function of the state in modern forms of subjectification (or the making of the social bond)—it simply becomes one of an array of mechanisms and apparatuses. As Nicolas Rose and Paul Rabinow put it in their gloss of Foucault's governmentality:

> We must now investigate the powers of the state and of its apparatus of rule in relation to all those many transactions where our own concerns

37 Barbara Spackman translates Žižek's theory of the subject to the textual realm when she writes of the paradoxical binding of textual knowledge and textual nonknowledge as the locus of ideological fantasy in the text (*Fascist Virilities* xi).

> with our own lives have also become the concerns of others—not just explicitly political agencies, but also all those other authorities (religious, medical, commercial, therapeutic) who *whisper in our ears* and advise us how to act and who to be. [...] And, in a way that is disturbing to many, we can now recognize that the precepts, norms, and values disseminated in these practices of government have made us the kinds of persons we take ourselves to be. (emphasis mine; xi)

The nationalization of Italians as a biopolitical project was not the well-organized plot of a conniving and malevolent state (nor, for that matter, a benevolent one), as much as it was a varied and dispersed ensemble of technological engagements (colonial regimes of power-knowledge; medical, aeronautical, and cinematic innovation), textual enunciations, and techniques of visualization, which aligned themselves with a range of ideological positions (from "left" to "right," from "patriotic humanist" to "proto-fascist imperialist"), "whisper[ed] in the ears" of Italians in order to produce would-be modern—and, critically, racialized—political subjects.

Žižek's notion of ideological fantasy is key, insofar as the racial, biopolitical, and colonial discourses in the texts under discussion are structured by fetishistic logics of disavowal. Like Stewart-Steinberg's *Pinocchio Effect*, "a mode of thought that seeks to negotiate between the anxiety about the potential emptiness regarding national existence and the bond that nonetheless ties that Italian to his or her national existence" (367), the texts under consideration in this book "know" very well that the Italian national body and its "mutilated" or "lost" territories (Fiume, Libya, etc.) were never unified or whole to begin with, but all the same they "behave" as if they were. The rhetorical role of territorial and corporeal loss—from colonial military defeat and emigration to physiological degeneration and dismemberment—which I argue pervaded the Italian biopolitical discourse under consideration functions fetishistically: by recalling the loss of Italian emigrants, as Crispi did at this chapter's outset, the "missing" limbs of Italian colonial territories, or the military defeat at Adwa in 1896, to name just a few examples, the fundamentally empty (to paraphrase Stewart-Steinberg), or constitutively fractured nature (to paraphrase Esposito) of Italian corporeal, territorial, and political identity is thus suspended. In this way, then, quite paradoxically, rhetorical *loss* serves to mend a constitutive *absence* in the unitary, racialized Italian subject.[38]

38 The fundamental lack that characterizes the Italian national subject was decried perhaps most memorably and explicitly by Giacomo Leopardi in his 1824 *Discorso sopra*

As this book calls for a reading of the rhetorical making of that vital subject in light of early Italian racial and colonial discourse, it bears noting that fetishism also binds the European nation-state to racism and its colonial encounters. William Pietz's ethno-historical reading of the fetish as it emerged as a result of sixteenth and seventeenth-century colonial Portuguese encounters with West African societies and Homi Bhabha's influential reading of the colonial stereotype as fetishistic thus also situate the primary structure of ideological fantasy within the postcolonial frame. Bhabha uncovers a structural link between the disavowal of sexual difference (Freud's fetishist), and the disavowal of racial difference that the stereotype (as fetish) enables (74). Žižek's ideological fantasy lends itself particularly well to the ambivalent fictions that inhere in both racial and colonial discourse.

The Race for Modernity: From Normal Man to Mechanized Post-Humanity

For readers familiar with the context of modern Italy, my discussion of racial discourse and biopolitics between Unification and World War I will likely bring to mind two of modern Italy's most provocative and infamous theorists of the relationship between human bodies, politics, and technology: criminologist Cesare Lombroso (1835–1909) and the founder of Futurism F.T. Marinetti (1876–1944). Indeed, in many ways, Lombroso and Marinetti might be said to constitute the representational bookends of the constellation of terms that this book assembles in order to map the principal coordinates of textual strategies of racialization in post-Unification Italy. Lombroso's relentless explorations of the physiological traces of criminality and degeneracy sought to prove the indexicality of the human body and the existence in nature of *homo criminalis*. In his two most famous studies, *L'uomo delinquente* (Criminal Man, 1876) and *La donna delinquente, la prostituta e la donna normale* (Criminal Woman, the Prostitute, and the Normal Woman 1893) and throughout his vast oeuvre, he deployed diverse technologies—among them photography, craniometry, and modes of writing—that have been linked to a distinctively modern rethinking of the (biological) subject's relationship to the social.[39]

lo stato presente dei costumi degli'italiani (Discourse on the Present State of the Customs of the Italians) (Leopardi, *Poesie e prose*).

39 For this reading of Lombroso, see Suzanne Stewart-Steinberg's chapter, "In a Dark Continent: Cesare Lombroso's Other Italy," in Stewart-Steinberg, *The Pinocchio Effect.*

If for Lombroso men and women were irrevocably bound to the limitations of their physiology, Marinetti envisioned a hyper-modern post-humanity capable of overcoming its organic and hence limited composition in favor of a mechanized and impenetrable man. As Jeffrey Schnapp recounts, "The overcoming of physical decay by forging new bodies and materials had always figured among futurism's heroic themes (and never without nationalist and/or imperialist connotations). [...] Futurism was deeply haunted by the problematic of decline, whether in the domain of nature, the individual body, or the body politic" (198). From Futurism's birthplace in the "maternal ditch" overflowing with the "nourishing sludge" that Marinetti recalls having suckled from the "blessed black breast of [his] Sudanese nurse," through its most infamous novelistic incarnation in *Mafarka, le futuriste* (whose controversial opening chapter, "The Rape of the Negresses," along with some of the novel's other more "colorful" elements, made Marinetti the subject of a scandalous and wonderfully opportune trial for public indecency upon its translation into Italian in 1910), and beyond, vividly aestheticized racial otherness was quite self-consciously thematized as an integral part of Futurist poetics. The Futurist response to nineteenth-century physiological and literary decadence was both a turning towards embodiment (through a poetics of pulsating and abject ebony bodies, the black breasts of Sudanese wet nurses) and its neutralization through a transformation of its organic limitations into objects of desire and conquest (Sartini Blum 84). Though I would hesitate to situate Lombroso and Marinetti unproblematically along a biopolitical continuum that begins with the former's crude scavenging for the corporeal traces of criminality and finds its fullest expression in Marinetti's rambunctious techno-imperialist poetics, it might be helpful to keep them in mind as placeholders along this trajectory of race, biopolitics, and Italian modernity. While the implications of these two thinkers and their respective works for Italy's history of racial representation remain to be fully explored, I have chosen to hold them in reserve for the present study, as their respective elaborations of racial difference—depicted with brash and unapologetically black and white brushstrokes—set them apart from the subtler, more insidious brands of racialization that emerge in the chapters that follow. Instead, this book focuses on texts across a range of fields whose racial investments are harder to pin down precisely because they, unlike those of Lombroso and Marinetti, are not at first glance so hysterically racist. The texts under consideration in this book are not populated with the kinds of racialized others and/or deviants on whom so much of these two thinkers' respective works rely.

Rather than attempting to enshrine a new canon of racist Italian literary or political thought, or attempting to detail the institutional or cultural bases for the emergence of the Italian racial discourse, this book examines an eclectic range of canonical and non-canonical narrative forms in order to sketch the discursive, logical, and visual productivity of race, before a historical and theoretical backdrop that is unequivocally biopolitical.[40] Stewart-Steinberg's thoughtfully researched book covers an impressive ground with her reading of the roots of Italian modernity in the (invisible) ideological "strings" that tie Pinocchio to post-Unification nationalism. *Vital Subjects* engages this and other critical readings of Italian modernity by contributing an analysis of the territorial and biological grounds on which Italian racial discourse sought to anchor itself in the years between Unification and World War I.[41] As Stewart-Steinberg herself concedes, her study of formulations of the social bond (namely, the project of "making Italians") in this period of Italian history does not attempt to account for the fact that race, particularly when tethered to the seductive rhetorics of affect and education that Stewart-Steinberg's study does trace, exerted a considerable influence on how those bonds were envisioned. What follows is by no means an exhaustive historical treatment of the vast field of racialist discourse that extended from political doctrines, through the (social) scientific disciplines, to national literature and popular knowledge. Such ambitious and wide-reaching intellectual histories of the period in question are currently under way, illustrating a growing sense of awareness in the field of Italian literary and cultural studies about the degree to which the mythology of *italianità* has been shaped by race.[42]

40 For the reader who seeks a comprehensive account of the emergence of the racialized population as a target of intervention in post-Unification Italy, a number of recent monographs published in English provide thorough historical accounts of the institutional and social formulation of the Italian population in the fields of public health campaigns, demography and childhood, and statistics. See, for example: Ipsen, *Italy in the Age of Pinocchio; Dictating Demography*; Snowden, *The Conquest of Malaria*; *Naples in the Time of Cholera, 1884–1911*; Patriarca, *Numbers and Nationhood.*

41 In addition to the works I have already discussed, the readings of Italian modernity to which I refer are: Campbell, "'Infinite Remoteness'"; Fuller, *Moderns Abroad*; Horn, *Social Bodies*; Schnapp.

42 See Re, "Italians and the Invention of Race." Re's article, a thoroughly researched piece (totaling over fifty pages in length), argues that the Italian invasion of Libya marked a turning point in the history of racialization in Italy, as Italians turned what had been until then an internal racialization (of women, criminals, southerners, etc.) toward the outside (the Libyan "other"). Re suggests that Italy's racializing tradition has fundamentally literary origins. See also Giuliani and Lombardi-Diop.

CHAPTER ONE

Colonial (Re)productivity

> Colonialism was not a secure hegemonic bourgeois project. It was only partly an effort to import cultured sensibilities to the colonies but as much about the *making* of them.
>
> Ann Laura Stoler, *Race and the Education of Desire* (99)

One of liberal Italy's first and most influential proponents of demographic colonialism in the Horn of Africa, Leopoldo Franchetti (1847–1917), spent the earlier part of his career traveling Italy's southern regions on horseback, armed with rifles and intent upon, to modify Christopher Miller's phrase, "reaching out to the most unknown part of the [nation-state] and bringing it back as language."[1] Before beginning his career in parliament in 1882, Franchetti published two proto-sociological inquiries on Italy's southern regions that, along with Pasquale Villari's *Lettere meridionali* (Southern Letters, 1875), are generally considered to have inaugurated modern Italy's *questione meridionale* (southern question): *Condizioni economiche e amministrative delle provincie napoletane* (Economic and Administrative Conditions of the Neapolitan Provinces, 1873–1874), and *La Sicilia nel 1876* (Sicily in 1876) (which Franchetti co-wrote with Sidney Sonnino, who would later become Prime Minister).[2] The southern question was arguably the leading

1 Miller describes Africanist discourse as emerging from a European "gesture of reaching out to the most unknown part of the world and bringing it back as language" (5).
2 *La Sicilia nel 1876* included Franchetti's *Condizioni politiche e amministrative della Sicilia* and Sonnino's *I contadini in Sicilia*. It was later republished as *Inchiesta in Sicilia* in 1974. In addition to these well-known texts, I refer to "Sulle condizioni dei lavoratori agricoli," "Relazione alla Commissione reale pei demani comunali nelle province del Mezzogiorno," and "Mezzo secolo di Unità nell'Italia meridionale," which, along with *Condizioni economiche e amministrative delle provincie napoletane* and *Appunti di viaggio*

question for policymakers after Unification. Villari and Franchetti were two pioneers of what became known as *meridionalista* (southernist) literature—a proto-sociological genre which depicted southerners as either hapless victims of history plagued by poverty, illiteracy, superstition, and other forms of "backwardness," or as criminally, culturally, and/or racially resistant to the rationalist progress of liberal-capitalist development. Franchetti's post in the Italian parliament and his engagement with Italy's southern question brought him to Italy's first colony, Eritrea, in 1889 (before it was officially designated as such) to conduct agricultural experiments and to advocate for the relocation of several dozen primarily southern Italian peasant families to expropriated land in the Eritrean highlands.[3] He envisioned that the Italian state might correct the economic and political injustices to which it had been subjecting its own southern peasantry by redirecting the increasing flow of its emigrating masses from *oltreoceano* (across the ocean) to the promised lands of its *oltremare* (across the sea). Named by Prime Minister Francesco Crispi as Deputy of the Special Mission to Colonize Eritrea in 1890, the statesman's "almost maniacal" support for Eritrean colonization prompted him to roam far and wide conducting surveys and experiments with what Paolo Pezzino alleges was "the same passion that he had brought to his earlier travels in the *Mezzogiorno*" (68). Between 1890 and 1914, Franchetti delivered a series of reports to parliamentary committees and state ministries on the status of both his Eritrean plan and, after the Italian defeat at Adwa, meditations on Italy's colonial venture in Libya.[4]

To be sure, Franchetti's design for a resolution to the southern question through east African colonization was not necessarily unique; his interventions were part of a larger debate about the benefits ("bloodletting") or detriment ("hemorrhage") of emigration and, beginning in the 1880s, the potential presented by east African colonization. In the wake of an

are published in Franchetti, *Mezzogiorno e colonie* 1–258. Unless otherwise noted, all translations are my own.

3 Eritrea was produced (territorially and juridically) by Francesco Crispi, who named it in 1890. Franchetti's title was "Deputato in missione speciale per la colonizzazione dell'Eritrea." Del Boca, *Gli italiani in Africa Orientale* 384. See also Choate, *Emigrant Nation* 34.

4 I refer to the following texts: "L'Italia e la sua Colonia Africana" (1891), "Relazione sull'operato dell'Ufficio di Agricoltura e Colonizzazione dell'Eritrea" (1894), "L'avvenire della nostra Colonia" (1895) and "L'Italia e le sue colonie" (1914), which are published in Franchetti, *Mezzogiorno e colonie* 258–491. Unless otherwise noted, all translations are my own.

agrarian crisis to which Franchetti's *Condizioni economiche e amministrative delle provincie napoletane* and *La Sicilia nel 1876* were a direct response, there was a dramatic shift in patterns of Italian emigration (Romano and Vivanti 1750).[5] If the contours of previous emigration had been shaped by a combination of small business owners, artisans, and agricultural laborers from disparate Italian regions (including Sardinia, Sicily, the Italian Alps, and the Ligurian Riviera) who sought seasonal employment in Europe or North Africa, emigration began to take on a different character after 1880: significantly larger numbers of peasants, primarily from the continental South, began to leave Italy—often definitively—for the Americas.[6] A concomitant demographic explosion meant that the Italian state was also concerned with balancing an increasingly prolific population with economic development that would curb what it identified as the consequent threats of brigandage and the fomentation of socialist resistance to the practices of the liberal state.[7]

In 1874 alone, two years before the Italian state began collecting emigration statistics, at least four substantial volumes on the emigration and unofficial colonies of Italians abroad (in Europe, the Americas, Africa, and the Levant) were published and reviewed in the popular journal *Nuova*

5 See also: Choate, *Emigrant Nation*; Gabaccia.

6 Historian Frank Snowden has called for a revaluation of the role that public health played in liberal Italian emigration, by illustrating that malaria was one of its primary causes. Snowden writes, "Here, claimed the antimalarial crusaders, was a major cause of the massive hemorrhage of the most able-bodied and hardworking southern male youths who migrated abroad by the millions between 1880 and the First World War. There they hoped to build the economies of Italy's competitors instead of developing productivity at home" (Snowden, *The Conquest of Malaria* 21).

7 Tekeste Negash writes of a "violent population explosion" (15) and claims that between 1861 and 1911, the Italian population increased from 25 million to 35 million and that emigration consequently increased from 1887 onwards. In his *Lettere meridionali*, Pasquale Villari, to whom Franchetti pays homage in his *Condizioni politiche e amministrative della Sicilia* for having been the first to articulate the *questione meridionale*, employs, significantly, a metaphor of sickness to describe the socialist threat that the landless southern peasantry presented: "The most dangerous sickness of modern societies [...] is socialism. [...] We must think about it before the multitudes do" (cited in Romano and Vivanti 1714). Franchetti is haunted by the spectre of socialism when he writes of the rural masses: "Now, the rural southern plebes are having a hard time organizing themselves. Once they do, they will become a dangerous explosive that blows up in the hand of whomever attempts to manage them" (*Mezzogiorno e colonie* 223). See also the chapter "A World at War: The Italian Army and Brigandage" on *brigantaggio* as a textual construction in post-unification Italy in Dickie, *Darkest Italy* 25–53.

antologia (Girolamo; Choate, *Emigrant Nation* 25). Contributors decried the distinguishing feature of Italian emigration in the 1870s as the "nearly absolute deficiency of capital." (Girolamo 627) Emigrants were considered "*braccia*" (arms) leaving the nation, and observers noted that, in the case of the thousands of Italians bound for Buenos Aires in 1870, a third were bound for *asilo pei poveri* (homeless shelters); numerous others endured "cruel suffering" from New York to Boston (627). During the first parliamentary debates of emigration in 1888, Franchetti and his fellow policymakers Giustino Fortunato, Andrea Costa, and others discussed the potential benefits of restricting emigration: was emigration a necessity, an "inevitable evil"? Would limiting it increase Italy's population and suppress wages? Could internal emigration be encouraged as a solution to emigration abroad (Choate, *Emigrant Nation* 28–29)? Was it was possible to turn emigration from a phenomenon of poverty and backwardness into a colonial politics of vigor and productivity?

In the midst of increasing economic and political instability in the South and the consequent exodus of hundreds of thousands of landless peasants, the Italian colonial defeat at Dogali in 1887—another of the fetishistic 'losses' recurring throughout liberal-era Italian colonial discourse—marked a significant moment in both administrative and popular imaginaries.[8] In January 1887, an army led by Ethiopian Ras Alula defeated a contingent of roughly five hundred Italian soldiers led by lieutenant colonel Tommaso De Cristoforis. The defeat was immediately commemorated in Italy. A monument to the fallen soldiers was promptly erected at the entrance to Rome's new train station and the piazza in which it was situated was renamed Piazzale dei Cinquecento.[9] In 1889, the prolific nationalist and imperialist commentator and novelist Alfredo Oriani published a commercially successful collection of essays entitled *Fino A Dogali* in which he exalted the dead soldiers as heroes and calls for Italy's perseverance in the conquest of Africa. Eighteen eighty-seven also marked Francesco Crispi's first election to the office of Prime Minister.[10] Crispi worked closely with

8 For more on the Battle of Dogali and its significance, see Del Boca, "Realtà e leggenda di Dogali."

9 See von Henneberg for a description of the monument, as well as the implications of its strategic positioning and repositioning in the 1920s.

10 The first year of Crispi's term as Prime Minister was marked by military defeat at Dogali, and the collapse of Crispi's last government was triggered by another, more significant colonial defeat at Adwa in 1896.

Franchetti and was initially an enthusiastic proponent of Franchetti's plan to resolve the southern question and curb the depopulation of Italy's southern countryside through demographic colonialism in the Horn of Africa. In 1890, Crispi joined Asmara and the ports of Assab and Massawa to form the first Italian colony of Eritrea. As historian Mark Choate describes Crispi's project (paraphrasing Franchetti himself), "Territorial settlements would strengthen Italy's African colony and allow masses of emigrants to thrive amid transplanted Italian customs, traditions, and society in the shadow of the Italian flag" (*Emigrant Nation* 32).

In the post-Unification years, the national politics of the southern question and mass transatlantic emigration brought neglected Italian populations into the realm of visibility for policymakers. Franchetti's work illustrates how this project culminated in a distinctively Italian brand of colonialism. Italy's demographic colonialism in the Horn of Africa, which was largely directed at Italians themselves rather than local populations, was thus linked to a wider array of biopolitical concerns and practices aimed at the guiding, or "making" of Italians as *vital subjects* in Italy and abroad. Indeed, as Franchetti was continuing to advocate for Italian colonialism in Africa, he and his wife Alice Hallgarten (1874–1911) founded and directed a school for agricultural workers and their children on his private Villa Montesca in the Umbrian Città di Castello (Perugia). This chapter therefore brings together Franchetti's writings on the southern question, early colonialism, and his pedagogical project at the Villa Montesca in order to explore the breadth of what I call his *colonial biopolitics*. What follows is a close reading of the discursive and ideological links among a range of colonial forms and practices—from southern Italy to Eritrea and eventually the Umbrian countryside—employed by this influential thinker. His writings have been of interest almost exclusively to historians whose goals range from reconstructing the cultural, political, and ideological bases for debates around Italy's southern question and early colonial projects to tracing the breadth or recurrence of his stereotypical representations of the Italian South.[11] Recent scholarship on the Villa at Montesca has recast "Baron Franchetti" as a beneficent patron of secular pedagogy, given his and his wife's hosting there of Maria Montessori as she penned her treatise that would revolutionize early childhood education. The splitting of Franchetti's oeuvre by scholars has produced some ambivalent results, as Franchetti emerges as either a dealer

11 See: Moe, *The View From Vesuvius*; Dickie, *Darkest Italy*; Wong; and Choate, *Emigrant Nation*.

in quasi-racist stereotypes, a clear-minded liberal pragmatist, or a generous philanthropist and experimental pedagogue.

Franchetti figures so-called "demographic" colonialism in the Horn of Africa as an answer to what he suggests is an improper circulation of bodies and labor—namely, Italian emigrants laboring under the shadow of another nation-state's flag. This chapter reads the three fields of Franchetti's social intervention alongside one another in order to suggest that his answer to the problem of the improper circulation of laboring Italians as a result of emigration was formulated as an agricultural project the rhetoric of which overlapped with a biological imperative. Rather than continuing to set out for foreign lands, according to Franchetti's vision, prolific Italian bodies were to cultivate prolific Italian colonies, both at home and abroad. Biopolitical discourse in post-Unification Italy was acutely aware of its constitutive fractures—of both the body politic (poor, landless emigrants; illiterate peasants; lawless southern "brigands," "mafiosi," and so on) and the lands it either inhabited ("America") or sought to (longed-for colonies in Mediterranean and East Africa). An ideological fantasy, the "making of Italians" as a unified corporeal body was nourished by rhetorical figures of loss, fragmentation, and dismemberment. By insisting on loss (from emigration, colonial defeat, etc.), rather than absence, turn-of-the-century Italian biopolitical discourse was able to disavow the constitutively fractured nature of that body. In calling forth this cultural preoccupation with corporeal and territorial fragmentation, Franchetti's brand of demographic colonialism aimed at least as much to make Italians themselves 'whole' as it did to colonize African others. His texts therefore also illustrate what some scholars of Italian colonialism have called the "peculiarity of the Italian 'civilizing mission,' which was directed as much toward Italian colonizers as it was to the colonized" (Ben-Ghiat and Fuller 3). As Franchetti would recall to his colleague at the Associazione per gli Interessi del Mezzogiorno (Association for the Interests of the South) and future biographer Umberto Zanotti-Bianco, the years after Unification were characterized by the fear that "with every little uprising in Italy, with every international situation, it seemed we were seeing the miraculous work of our unification wrecked" (Pezzino 16). Tantamount to a "miracle," Unification was thus fraught with anxiety about the fragility of the national project.

Love thy Nation: Affect and Rule in Franchetti's South

> It is our sacred duty [...] to commemorate him, [one] of the most noble intellects that new Italy possessed, [and] to offer him a profound goodbye, a salute which [I] am honored to offer in the name of the provinces of southern Italy, which Leopoldo Franchetti knew and loved like none of us knew and loved.
>
> Giustino Fortunato, *Leopoldo Franchetti. Ricordi* (x)

The goal of Franchetti's travels through mainland southern Italy and Sicily was to produce a uniquely Italian *savoir*, or knowledge, about the Italian South through objective fieldwork:

> In Italia, chi voglia imparare a conoscere le condizioni del paese, pur troppo così poco conosciute, e ricercare i suoi bisogni e i rimedi dei suoi mali, non deve contentarsi di studiar nei libri, quasi tutti forestieri, l'economia politica, l'amministrazione o il diritto costituzionale ma terminati gli studi teorici, si alzi, cinga i lombi e vada a vedere coi propri occhi, a sentire colle proprie orecchie, vada a constatare i fatti, e a verificare se giustifichino le teorie degli scrittori. (Franchetti, *Mezzogiorno e colonie* 53)
>
> [In Italy, whomever wants to learn to get to know the conditions of his country (which are unfortunately quite unknown), to research its needs and the cures for its ills, should not content himself with studying books (almost all of which are written by outsiders), nor its political economy, government administration, or constitutional law. Instead, once he has completed his theoretical studies, he must stand up, gird his loins, and go see with his own eyes, hear with his own ears, to ascertain the facts, to verify whether they justify the theories.]

Within Franchetti's purportedly objective (and explicitly gendered) disposition toward the representation of the Italian South lies the specter of colonial discourse. Homi Bhabha's description is particularly apt in describing Franchetti's approach to southern Italy:

> [Colonial discourse] produces the colonized as a social reality which is at once an 'other,' and yet entirely knowable and visible. It resembles a form

> of narrative whereby [...] subjects and signs are bound in a reformed and recognizable totality. It employs a system of representation, a regime of truth, that is structurally similar to realism. (70–71)

Franchetti's social realist texts contain just such a tension between southern alterity, on the one hand, and its fully visible representability.

Yet despite Franchetti's explicit claim to objectivity, among the most striking discursive strategies of his 'realist' southern texts is a recurrent vocabulary of love and desire. Franchetti, who positions himself as both an objective social explorer and a foreign traveler, repeatedly justifies his project by claiming that it has sprung forth from a love of both social science and of the abject populations that he takes as his object.[12] Concluding an 1883 report "On the Conditions of Agricultural Workers," Franchetti writes, "I would like to ask your permission to say a few more words about another wrong done to the southern peasantry, something that has weighed on my heart for a number of years, and from which I have learned to know and to love the Italian South" (*Mezzogiorno e colonie* 165). Franchetti employs this strategy in part in order to manage his status as *forestiero* (outsider). By deploying a rhetoric of amorous compassion and charity, his texts attempt to stave off (what he perceives as potential) objections to a Tuscan taking up pen (and rifle) against southern populations.[13] He opens his *Condizioni economiche e amminstrative delle provincie napoletane* with the following dedication:

> (S)arei profondamente addolorato se potessero nelle persone sincere di quella parte d'Italia, eccitare quel patriottismo locale male inteso, che

12 For a reading of *Condizioni politiche e amministrative della Sicilia* that highlights Franchetti's use of rhetorics of surprise, shock, and conventions of the travel narrative, see Moe, *The View from Vesuvius* 241–244.

13 Moe also discusses this passage, emphasizing Franchetti's avowed unifying rhetoric. See Moe, *The View from Vesuvius* 238. In his oft-quoted opening to *Condizioni politiche e amministrative della Sicilia* (his portion of *La Sicilia nel 1876*, later republished as *Inchiesta in Sicilia*), Franchetti describes how his initial impressions of Sicily's kind people, abundant citrus groves, and picturesque landscapes with time give way to a much bleaker and more violent picture of the island, as he mentions the "profound tenderness" he begins to feel for his rifle. The "paradise inhabited by devils" topos is ubiquitous in literature on Sicily and the southern regions more generally. In his recent history of the Sicilian mafia, John Dickie confirms that, on their voyage to Sicily, Franchetti and Sonnino were indeed armed with "repeating rifles and large-caliber pistols" (Dickie, *Cosa Nostra* 54–60). See also Moe, *The View from Vesuvius* 242.

> nega tutto e rifiuta di cercare i rimedi ai mali piuttostochè convenire con un forestiero di cose che tornino a disdoro della sua regione, provincia o comune. Siamo tutti Italiani, le loro vergogne sono nostre, siamo deboli della loro debolezza. (Franchetti, *Mezzogiorno e colonie* 3)
>
> [I would be deeply hurt if [my writings were to] excite that misunderstood local patriotism [...] that denies everything and refuses to try to find remedies instead of agreeing with an outsider about [...] their region, province, or commune. We are all Italians, their disgraces are our own, and we are weakened by their weakness.]

Franchetti adopts a stance of humble objectivity to justify his study by appealing to a nationalist rhetoric of equality and commonality. Yet, by following "we are all Italians" with "we are weakened by their weaknesses," Franchetti exposes the tentativeness of this initial gesture with an "us" and "them" equation that evokes the threat of contagion.[14] Indeed, this formulation exposes the immunitary structure of the newborn Italian national community. This opening passage also establishes what will be a recurrent mode of representation within Franchetti's southern texts, as it ties affect to the production of knowledge, and thus, to technologies of rule: Franchetti-as-narrator will be "profoundly hurt" if his status as "outsider" inhibits his readers from drawing administrative strategies and practices from his texts. Since 1860, Franchetti claims, the Italian state had neglected the peasantry, as lawmakers had "lost sight of" the agricultural class (Franchetti, *Mezzogiorno e colonie* 158). Accordingly, his texts aim to restore the visibility of peasants to state administrators. As Bhabha argues, this visibility was far from neutral, tied as it was to the practices of disciplining and regulating difference enacted by the newly unified Italian state and its institutions.

Franchetti's ideal administrator is a figure that his texts repeatedly attempt to sketch. In *La Sicilia nel 1876*, his main objective is to present information about the current systems of rule and to establish and condemn the Sicilians' mode of self-governance by juxtaposing their adherence to the private rule of violence to the liberal state's public rule of right and law, in order to inform the state's appropriation of them.[15] Franchetti's exemplary southern administrator

14 For an alternative reading of the same passage which emphasizes Franchetti's attempt to "nationalize the problems of the south, dissolving the force of regional difference in an equation of theirs is ours, they are us," see: Moe, *The View from Vesuvius* 238.

15 The canonical literary interpretation of the encounter between the Sicilian

is necessarily from northern Italy (a point to which I will return) and his main charge is to effect the political, economic, and moral education of southern political subjects. This educative campaign is represented in Franchetti's texts as a battle that is to be waged at the level of sentiment. Southern deficiency is defined primarily in terms of a lack of affection for the state: ("they do not feel a social sentiment," Franchetti and Sonnino 133; "that missing sentiment for the law and for legal security," 141; "they have maintained the dynastic and superstitious affection for the Bourbons," Franchetti, *Mezzogiorno e colonie* 28, etc.) and a good southern administrator is characterized by an excess of affect and desire for the rule of law:

> Allora solamente sarà possibile trovare i modi di dare all'intera amministrazione civile e giudiziaria, uniformità nello spirito e nell'indirizzo, e *di infondere in tutti un sentimento* tale che [...] ogni impiegato [...] intenda il fine comune al quale dovrebbe esser diretta l'opera dei singoli funzionari, e provi per esso *quell'amore del quale ogni uomo intelligente si sente preso* [...] per uno scopo grande e difficile. (Franchetti and Sonnino 266)
>
> [Only then will it be possible to find ways to give the entire civil and legislative administration a uniformity of spirit and application, *and to instill in everyone such a feeling* that [...] every clerk understands the common goal toward which his individual work should be aimed and feels for this goal *that love that captivates all intelligent men*.] (emphasis mine)

And:

> (E)ntrando in quelle provincie, s'aspetterebbe a trovare un Eden politico ed amministrativo, una classe dirigente che, acquistati coll'uso dell'autorità *il sentimento della responsabilità e della dignità,* le tradizioni amministrative e *l'amore alle cose pubbliche,* governi ed educhi una popolazione docile, *più coll'amore e colla fiducia che coll'autorità,* e la prepari gradatamente ad entrare a fare parte del governo. (Franchetti, *Mezzogiorno e colonie* 25)
>
> [Entering those provinces [Abruzzo and Molise], one might expect to find a political and administrative Eden, or a leading class that, having

aristocracy and representatives of the newly unified Italian state is in the meeting between Don Fabrizio of Salina and Chevalley di Monterzuolo in Tomasi di Lampedusa's novel *Il gattopardo* (*The Leopard*).

> assumed with *authority a feeling of responsibility and dignity [and] a love for public service*, would govern and educate a *docile population with love and trust rather than with authority*, and would prepare it to enter into government.] (emphasis mine)

Thus, southern subjects—a "docile population"—and the administrators that are apt to govern them are bound to one another by a rhetoric of affect. As Bhabha has suggested, colonial discourse is marked by both pseudo-objective modes of knowledge production—"learning," "discovery"—and by subjective desire and pleasure—"dreams," "fantasies" (71). Franchetti's southern texts are steeped in such an oscillation: Franchetti emerges as a colonial narrator as he presents himself as compelled by his love for his southern objects and his desire for them to confront the horror of their disorder and to represent it scientifically. Furthermore, the binding of the compassionate administrator to his unfeeling southern subjects in an inverse relationship (lack/abundance of sentiment) illustrates how a rhetoric of sentiment can be actively mobilized to legitimize subjection and rule. For Bhabha, this ambivalent binding of sentiment and subjection, of objective knowledge and subjective affect, marks all colonial discourse as fundamentally fetishistic. As he puts it, "This conflict of pleasure/unpleasure, mastery/defence, knowledge/disavowal, absence/presence, has a fundamental significance for colonial discourse. For the scene of fetishism is also the scene of the reactivation and repetition of primal fantasy—the subject's desire for a pure origin that is always threatened by its division [...]" (75). This oscillation also pervades Franchetti's immunitary logic of "us" and "them" ("We are all Italians, their disgraces are our own, and we are weakened by their weakness," cited above), which fetishistically binds and splits the national population in two.

Franchetti's Clash of Civilizations

Franchetti's southern texts grapple with the possibilities of naturalized difference within the borders of the newly forged nation-state. His texts are to be both descriptive and prescriptive, as he illustrates structurally by organizing his inquiries under subheadings such as "Conditions" and "Remedies." Franchetti's writings may be situated within the context of what David Horn has identified as a flurry of intellectual activity beginning in the nineteenth century that forged the social sciences and created "social bodies" as their object—men and women, "located neither 'in nature' nor in the private sphere, but in that modern domain of knowledge and intervention carved out by statistics, sociology, social hygiene, and social work" (*Social Bodies* 4). This moment of negotiation—in which boundaries between public and private are being rethought through the analytic of the social—emerges in Franchetti's southern texts as a tension between whether his nightmare of moral and economic degeneration within the boundaries of the nation-state can be explained through socio-historical forces such as foreign occupation and oppression, economic systems, the organization of labor-power and private property, or through biology, that is, naturalized or pathologized difference figured as a threat to the social body.

Despite Franchetti's avowed subscription to social determinism, his southern texts betray an ambivalence about to how to represent southern populations. These populations are unequivocally marked with difference, and thus as targets of this emergent mode of social intervention, but Franchetti's texts betray a murkiness with regard to precisely where socio-historical forces cease to be sufficient in explaining the economic and moral disparities that his project so desperately seeks to expose and ultimately level. At these points in the texts, causality is fragile and ambiguous. For instance, Bourbon occupation is invoked as explanatory, but Franchetti's furious repetition of the entrenchment of its power effects within the psyches of the southern populations over thousands of years seems to point more to an evolutionist model of innate psychology (*Mezzogiorno e colonie* 219).

Franchetti's southern texts painstakingly construct several criminal elements of the population—*mafiosi, malfattori, briganti*—that threaten the body politic. In isolating these groups and prescribing their elimination, he adopts a rhetoric of sickness. This was a recurring discursive strategy in meridionalist literature after Unification, as the South and its inhabitants were referred to variously as "gangrene," a "bloody plague," or a "wound." In Franchetti's texts, examples abound of dangerous populations that are diseased, that infest the landscape and the population and are potentially contagious. In one instance, the *mafia* is positioned in opposition to the state as an infectious agent: "[I]f the Government does not want to itself undergo contagion by the conditions of the Island instead of curing it, if it does not want to become yet another Sicilian mafia, it may only govern it by the force of the modern State" (Wong 20). In *La Sicilia nel 1876*, such a rhetoric enables Franchetti's final prescription for the relocation of "the Sicilian element" to the mainland.[16] In his section aptly

16 The expression recurs several times. See also: Franchetti, *Mezzogiorno e colonie* 156; 167; 203.

entitled, "Remedies," the sickness to be cured is grafted explicitly onto Sicilian bodies, rather than onto their amoral practices or onto the geopolitical space of Sicily (as in the remainder of the text). Sicilians are thus to be excluded from self-governance, as they are wholly incapable of enunciating their political and economic needs within the framework of the liberal state. In a rare moment of carefully elaborated narrative, Franchetti writes:

> Spesso il sentir l'ammalato lamentarsi della sete, è pel medico una ragione per non dargli da bere. Spesso le sensazioni di cui l'ammalato si lamenta più aspramente, sono segno pel medico che i suoi rimedi sono efficaci e portano la guarigione. Spesso un sollievo momentaneo ed un miglioramento apparente è segno che il morbo peggiora, e la morte è vicina. (Franchetti and Sonnino 221)
>
> [Often, hearing the patient complain about thirst is, for the doctor, a reason not to give him something to drink. Often, the sensations about which the patient complains most bitterly are signs for the doctor that his remedies are effective and healing. Often, a momentary relief and an evident improvement is a sign that the disease is worsening and that death is near.]

Thus, the Italian state emerges as the doctor that can and must diagnose, treat, and cure a dying patient who can no longer articulate his own interests.[17] It is within Franchetti's deployment of sickness in representing southern populations that the ambiguity between social and biological causality that I mentioned above emerges.[18] According to Franchetti's logic, healthy bodies are those deemed capable of economic, moral, and political redemption through social intervention. Yet, the moment these bodies are deemed diseased, they are figured as contagious and necessitate containment or expulsion. Socio-historical forces are figured as almost physiologically

17 See also: Moe, *The View from Vesuvius* 246–247.

18 In her chapter, "Fascism as Discursive Regime," Barbara Spackman proposes that, "Fascism's principal fantasy was a reproductive one." Through a careful analysis of Mussolini's "Discorso dell'ascensione," Spackman traces a collapsing of biological and social reproduction within Mussolini's call for "igiene sociale, profilassi nazionale" and locates fascist racism within this conflation. She writes, "[T]he regime's policing of gender and sexuality, its codification of that policing in pronatalist policies and reproductive incentives and controls, was no afterthought but a part of the very formation of fascist ideology" (Spackman, *Fascist Virilities* 144).

corrosive. He writes, for instance, "The [peasants'] mentality [is] that of the historical period in which the town [is] arrested" (*Mezzogiorno e colonie* 219). Years of foreign occupation and/or self rule have arrested the development of southern minds and bodies and rendered them, like strains of an infectious disease, resistant to intervention and change. Socio-historical causality slides into an almost naturalized, and unequivocally colonialist, distinction between modern and savage and rational and irrational:

> Alla vista di quella desolazione, il forestiero è tentato di credere che in quel paese, ogni anno dopo il raccolto, avvenga qualche grande sciagura, qualche invasione, qualche conquista che tolga i frutti di tutto il lavoro dell'anno [...] oppure che da secoli e secoli i raccolti cattivi si siano seguiti senza tregua ed abbiano appena lasciato ai proprietari ed ai lavoranti tanto da poter mangiare e seminare, oppure che in quel paese viva una qualità di uomini speciale, che, in mezzo a terre coltivate, abbia conservato l'imprevidenza dei selvaggi delle praterie d'America [...] che [...] non senta quel desiderio comune a tutti gli uomini di migliorare la propria sorte. (60)
>
> [Seeing such desolation, the foreigner is tempted to believe that in that village, each year after the harvest, there must be a great disaster, some invasion or conquest that steals the fruits of the entire year's work [...] or else that for centuries and centuries there had been one bad harvest after another, with no respite, and that it had left proprietors and workers just enough to eat and sow, or else that in that village there must live a special type of man who, in the middle of cultivated lands, had maintained the improvidence of American savages, and who does not feel that desire common to all men to improve his own lot.]

Franchetti skates the line between naturalized and social differentiation: atavistic remnants of a pre-modern past prohibit the entrance of these elements of the population into humanity by inhibiting the evolution of "that desire common to all men to improve his own lot." What distinguishes this "special type of man"—the southern agricultural laborer—from other, ostensibly less troubling types is his inability to act in his own self-interest. Quite critically, this difference is articulated through an assertion of the common: what is "common to all men" (a "desire [...] to improve his own lot") is also what divides them into types.

In the final instance, Franchetti's analysis falls back upon a distinction between two opposing *civiltà*. Franchetti's use of the term *civiltà* further underscores a nebulousness between naturalized and socio-historical difference: connoting either "culture" or "civilization," *civiltà* is an ambiguous signifier. For instance, *civiltà* and its correlate *civilizzazione* emerge in the writings of Giustiniano Nicolucci (1819–1904), founder of Italian ethnography and the man Maria Sophia Quine has called "the father of Italian racism," or the "Italian Gobineau," referring to genetically transmitted beliefs, customs, and values (127–152). In Franchetti's formulation, given that the term is shaped by a duality, *civiltà* implies a confrontation—more or less "civilized"—between the two groups or elements that it represents. Furthermore, Franchetti later employs the term in his colonial texts in order to distinguish between "past dominators" and Italian colonials: "We will demonstrate our civilization [*civiltà*] and our humanity far more than past dominators by imposing more mild and more stable tributes." (*Mezzogiorno e colonie* 271) The confrontation between Italian state and its southern subjects and territories is figured as a clash between civilizations in the following passage:

> La coesistenza della civiltà siciliana e di quella dell'Italia media e superiore in una medesima nazione, è incompatibile colla prosperità di questa nazione e, a lungo andare, anche colla sua esistenza, *poiché produce debolezza tale da esporla a andare in fascio al minimo urto datole di fuori. Una di queste due civiltà deve dunque sparire* in quelle sue parti che sono incompatibili coll'altra. (Franchetti and Sonnino 237)
>
> [The coexistence of the Sicilian civilization and that of middle and upper Italy in one single nation is incompatible with the prosperity of this nation and, in the long run, also with its existence, *insofar as it produces such a weakness that it exposes it to ruin with the slightest push from the outside. One of these civilizations must therefore disappear* with regard to the parts that are incompatible with the other [civilization].] (emphasis mine)

Franchetti bifurcates the nation-state into two antagonistic, almost naturalized populations and stages an external threat to divided, and therefore weakened, internal populations.[19] Furthermore, in this final passage, an analytic of war is explicitly evoked. Viewing such remarks through a biopolitical lens demonstrates how a discourse that isolates

19 For an alternative reading of the same passage focusing again on the logic of absolute difference, see Moe, *The View from Vesuvius* 245.

criminal "elements" for expulsion is informed by a discourse of race war (here figured as an inevitable clash between two ambiguous *civiltà*).[20] Moreover, an immunological model is already in place. The life ("prosperity" or "existence") of the nation is at stake, and vying for survival within it are two civilizations, one of which threatens to contaminate and weaken the other. In order to protect one, pathological elements of the other must "disappear." Nurturing, maintaining, and protecting life thus requires a certain kind of death. Writing about medical immunization, which requires the introduction of a small portion of the disease into the patient, Roberto Esposito suggests how it has become a generalizable social model, which he has named an "immunitary paradigm." "It's almost as if in order to save someone's life," he writes, "it is necessary to make them taste death" (*Terms of the Political* 61).

Salvaging the Productive Peasantry

What remains of that segment of the population—the landless peasantry—that Franchetti had deemed capable—indeed desperately in need—of moral, economic, and political regeneration? Franchetti's texts propose three possible outcomes for the landless peasantry if left within the borders of the nation-state: contagion by the criminal components of the population and descent into lawlessness ("The lifestyle of peasants is such that becoming a brigand is an improvement of their conditions rather than a worsening of them," *Mezzogiorno e colonie* 137); mass uprisings and peasant resistance to the liberal state ("The dissatisfaction of the peasants has manifested itself in various ways: brigandage, uprisings for the division of municipal property in the southern provinces, strikes, etc.," 155); and transoceanic emigration ("In 1872, 5,545 people emigrated from Basilicata, 5,150 of whom for America," 100). It is within this segment of the southern population that Franchetti most explicitly lays out a plan for social and biological engineering in the domains

20 Foucault writes, "The discourse of race struggle—which—when it first appeared and began to function in the seventeenth century, was essentially an instrument used in the struggles waged by decentered camps—will be recentered and will become the discourse of power itself. It will become the discourse of a centered, centralized, and centralizing power. It will become the discourse of a battle that has to be waged not between races, but by a race that is portrayed as the one true race, the race that holds power and is entitled to define the norm, and against those who deviate from that norm, against those who pose a threat to the biological heritage" (Foucault, *Society Must Be Defended* 61).

of sexuality, labor-power, and, ultimately, race through an ambiguous rhetoric of productivity. Franchetti eventually concludes that such a project should be rehearsed not at home, but within Italy's colonial territories in East Africa.

Franchetti singles out peasants as the most hard-working segment of the Italian population. Yet the failure of the Italian state to effectively harness this labor-power, claims Franchetti, has meant that the risks of criminal degeneracy, epidemics such as malaria, and sexual deviancy are explosive. In a striking passage, Franchetti conveys the horror of the behavior of southern peasants when not contained within the normalizing structure of the workweek:

> Questi contadini, forse i più laboriosi d'Italia, passano la domenica a giocare ed a ubriacarsi, e, al bisogno, si anneriscono la faccia, e vanno ad arrestare la gente per le strade maestre. Religiosi e superstiziosi al punto di spender migliaia di lire nei più poveri comuni per la festa del santo e per la fabbricazione della chiesa, non è raro sentirli parlar male dei preti. [...] Con tutto il loro rispetto pei signori, nelle sommosse reazionarie del 1860, sobillati da signori reazionari e da preti, assalirono le case dei signori liberali, e fecero morire uomini e donne in mezzo ai tormenti. I parricidi, i fratricidi sono relativamente numerosi. (*Mezzogiorno e colonie* 21)

> [These peasants, *perhaps the most hard-working people of Italy,* spend Sundays playing and getting drunk, and, when necessary, *they blacken their faces, and they go around harassing people on the main streets.* Religious and superstitious to the point of spending thousands of lire in the poorest communes for the feast of their saint and for the construction of their church, it isn't rare to hear them talking badly about their priests. [...] With all of their respect for the gentry, in the reactionary uprising of 1860, roused by reactionary gentlemen and by priests, they assailed the houses of the liberal gentleman and killed men and women in the midst of the torments. *Parricides, fratricides are relatively numerous.*] (emphasis mine)

Here, Franchetti suggests that the absence of disciplined labor results in a kind of racial and moral degeneracy: drunk southern peasants, having blackened their faces, are consumed by homicidal rage, which, in the worst instance, violently destroys even the family structure.

In another instance, concern with sexual promiscuity emerges from Franchetti's description of the extant agricultural labor structure in Italy's rural south:

> Il genere di vita dei braccianti d'ambo i sessi impiegati nelle masserie, è caratteristico, e più d'ogni altro, atto a dare un'idea dell'esistenza cui sono ridotti i contadini di quelle provincie [...] In alcune parti, dormono tutti e tutte nel medesimo stanzone; ogni famiglia vi si fa la sua lettiera di paglia a parte. Altrove vi sono due stanzoni, uno per le donne, l'altro per gli uomini, e il sorvegliante è incaricato del mantenimento dei buoni costumi; ciò non impedisce che, intorno a Matera, per esempio, il maggiore insulto che si possa fare ad una donna è il dirle: "sei stata alle masserie." (*Mezzogiorno e colonie* 84)
>
> [The lifestyle of the laborers of both sexes who are employed on the farms is characteristic more than any other, and apt to give an idea of the level of existence to which the peasants of those provinces are reduced. In some parts, they [male and female] all sleep in the same large room; every family makes its own separate straw bedding. Other places, there are two large rooms, one for the women, the other for the men, and the night watchman is charged with the maintenance of moral behavior; this does not impede the fact that, around Matera, for example, the worst insult that you can make to a woman is saying to her: "You've been in the *masserie*."]

Peasant labor conditions pose not only a moral threat, but a sexual threat as well. The site of moral, legitimate female labor (the *masseria*) is invoked as a euphemism for the amoral, illegitimate prostitute. Within Franchetti's southern texts, sexual relations are to be absorbed within the purview of the state. State management of labor-power is thus tied to a concern with productivity, in both senses.

Franchetti also marks the reshuffling of sexual relations that results from emigration as a target for intervention. Franchetti laments that one of the detrimental effects of permanent emigration is indeed the dissolution of the family (and class) structure, as large numbers of husbands abandon their wives in order to seek work abroad:

> Veramente, c'è molto di brutto nel modo in cui si opera l'emigrazione. [...] I costumi delle donne, nei paesi dove sono migliori, peggiorano. Le mogli degli emigrati finiscono spesso per cader sotto a qualche signore del paese, fanno figli in assenza del marito, e sono costrette spesso ad abbandonarli od a consegnarli, per essere portati all'ospizio dei trovatelli in Napoli [...] (*Mezzogiorno e colonie* 103)

> [There is truly a lot wrong with how emigration works. [...] The behavior of women, in the towns where it is optimal, gets worse. The wives of emigrants *often end up falling under some townsman,* they produce children in the absence of their husband, and they are often forced to abandon them or to turn them over, only to have them brought to the foundling home in Naples.] (emphasis mine)

The sexual threat alluded to here is hardly veiled by an economic one, as Franchetti fantasizes the abandoned peasant wife "falling under some townsman," thus conjuring up an image of both sexual and economic submission. Franchetti's texts betray a preoccupation with the biological proliferation of certain elements of the population. Children born to abandoned peasant wives clearly constitute another segment of the population that Franchetti's texts attempt to manage. These children would not only represent the refusal of sexual boundaries between economic classes, thus threatening their normative, hierarchical ordering, but, if abandoned, Franchetti warns, they would present a burden for emergent state technologies of social welfare. In his study of infant abandonment in nineteenth-century Bologna, David Kertzer notes that during this time Italy also witnessed an explosion in numbers of infants abandoned at foundling homes. According to Kertzer, adolescent male foundlings were often released for either military service or agricultural work. Females had a different fate: marriage or domestic servitude were the sole acceptable conditions for release, as home administrators were concerned with the potential for more reproduction out of wedlock, thereby ensuring a cyclical pattern wherein illegitimate babies would beget illegitimate babies.[21] In the above passage, Franchetti's nervous evocation of the foundling home is itself a site of reproduction: the biological reproduction of the abandoned peasant woman is thus tied to the social reproduction of the very conditions at which Franchetti's work is aimed, wherein such women give birth to either versions of themselves (abandoned girls), or to young men who would choose between spade and rifle. By the start of the twentieth century, Franchetti and his wife Alice Hallgarten would open their school at Villa la Montesca in order to provide job training and moral instruction to precisely these populations—orphans, delinquents, the underemployed, children of impoverished farmworkers, and the like.

21 For a fascinating discussion of the social regulation of prostitutes, midwives, wet nurses, and unwed mothers in late nineteenth-century Italy, see Stewart-Steinberg 223–228.

From Peasants to Soldiers: Franchetti's Hyper-productive Colony

By 1891, Franchetti had plotted a new course for the (re)productivity of southern agricultural workers. If their sexual relations were to be supervised and reproduction curbed within the confines of the nation-state, this was certainly not the case for Franchetti's Eritrean colonies. For Franchetti, Italy's Eritrean colony is "an absolutely new country" that is to be settled peacefully by hard-working Italian families. Franchetti occupies much of his colonial texts with topographical surveys and other "figural regimes," identifying territories that he believes are well suited for Italian settlement and cultivation (*Mezzogiorno e colonie* 313).[22] This requires the use of a common discursive strategy in colonial texts: a disavowal of the violent expropriation of lands from local populations that such settlement requires. Indeed, in the years Franchetti was writing, between 1893 and 1895, the Italian state conducted mass land expropriations, slating numerous tracts of locally owned land for habitation by future Italian settler-colonists. This resulted in frequent skirmishes between Eritrean peasants and Italian colonial authorities (Sòrgoni, *Parole e corpi* 21, 25). Franchetti makes no mention of this resistance, choosing instead to manage the obstacle that local inhabitants pose through a claim that vast amounts of uncultivated land already lay empty for the Italian taking. Furthermore, in actively producing the myth of *Italiani brava gente* (Italians as good colonizers), Franchetti outlines the ways in which Italian occupation will be kinder and gentler than past, non-European occupations and insists that local populations will be grateful to the Italian state for improving their living conditions.[23] He resolves the question of autochthonous populations with broad brushstrokes: "It is in the interest of the Government that the indigenous populations resume cultivation of the territories reserved for them, and that they return to the level of prosperity of which they are capable" (*Mezzogiorno e colonie* 296). Franchetti then moves on to delineate his project of Italian peasant settlement in great detail, making it clear that his concern lies more in colonizing Italians than local populations.

22 For a discussion of the "literariness" of "figural regimes," see: Hunt and Rudolf, and the Introduction to this book.

23 For more on the myth of *Italiani brava gente*, see Bidussa. Reflecting during World War I on Italian successes in recruiting Eritrean soldiers (*askari*) to fight alongside Italians against Ethiopian Emperor Menelik II, Franchetti comments that their fidelity "demonstrates that Italy [...] knows how to conquer the affections and trust of its indigenous subjects" (Franchetti, "Prefazione" vii).

In so doing, he illustrates the ways in which the management of the Italian population will be facilitated by relocating landless peasants and containing them within the colonies.

Franchetti stages the Italian colonial occupation of Eritrea as an opportunity for the Italian state to right historical wrongs inflicted upon its southern peasantry. If in his southern texts Franchetti painstakingly details the quantity and quality of uncultivated land within the boundaries of the Italian nation-state that is ripe for redistribution, in his 1891 report "Italy's African Colony," this opportunity is no longer figured as viable:

> [L]'Italia ha un debito da pagare verso le classi diseredate della fortuna, le quali col sangue e con le imposte hanno contribuito a far l'Italia, hanno contribuito alla occupazione africana; che è obbligo dello Stato italiano di far quanto sta in lui perché anch'esse possano guadagnarsi col lavoro, all'ombra della nostra bandiera, quella indipendenza economica inaccessibile ad esse sul suolo italiano. (*Mezzogiorno e colonie* 290)
>
> [Italy has a debt to pay to the classes who have been dispossessed of fortune, those who with their blood and their taxes contributed to making Italy and to African occupation; it is the Italian State's obligation to do all that it can so that even these [classes] can earn with work, under the shadow of our flag, that economic independence that is inaccessible to them on Italian soil.]

For Franchetti, Italian land is inaccessible to the hard-working peasants who populate it. Peasants are figured as worthy precisely insofar as they are soldiers, and as such have been drained both physiologically and economically (blood and taxes), both literally and figuratively, by the imbricated projects of "making" Italy and the colonial occupation of Africa. In addition, anxiety about transoceanic emigration is present from the outset, as Franchetti fantasizes accommodating these laboring settlers, "under the shadow of [the Italian] flag," rather than someone else's.[24]

24 A testament to the appeal of Franchetti's metaphor can be found in preeminent historian Christopher Seton-Watson's 1980 essay, where it is absorbed and presented as description: "The 'demographic imperialism' of southern politicians, publicists and peasants took the form of a search for land where Italy's surplus population could be settled in prosperity under the Italian flag" (170). While Seton-Watson lifts Franchetti's metaphor to describe his project, he is not cited explicitly in the article. The metaphor

Franchetti's writings illustrate the ease with which peasant populations might be contained and managed once relocated to colonial territories: his studies include countless figures and calculations of the nutritional intake, housing plans, labor conditions, and health of Italian laborers. In his "Report on Agrarian Experiments Executed in the Colony," Franchetti fetishistically records daily alimentary consumption of individual Italian workers:

> Ecco l'elenco, delle derrate che compongono il vitto giornaliero di ciascun operaio: Carne, Cg. 0,400 a lire 1, 34 il Cg. Pasta o riso, grammi 150 a lire 0,85 il Cg. Pane Cg. 1 oppure 1 Cg. di farina a lire 0,65 il Cg. Olio, cl. 15 a lire 2,50 il litro. Formaggio, grammi 10 a lire 4 il Cg. Conserva, grammi 10 a lire 1,40 il Cg. Cipolle, grammi 25 a lire 0,60 il Cg. Aglio, grammi 25 a lire 0,80 il Cg. Pepe, grammi 1 a lire 3 il Cg. Caffè, grammi 15 a lire 3,50 il Cg. Zucchero, grammi 22 a lire 1,20 il Cg. Rhum, cl. 4 a lire 3 il litro. Erbaggi in conserva, grammi 25 a lire 1 il Cg. (*Mezzogiorno e colonie* 342)
>
> [Here is the list of the foodstuffs that compose the daily provisions of each worker: Meat, .40 kgs. at 1.34 lire per kg. Pasta or rice, 150 grams at 0.85 lire per kg. Bread 1 kg. or 1 kg. of flour at 0.65 lire per kg. Oil, 15 cls. at 2.50 lire per liter. Cheese, 10 grams at 4 lire per kg. Preserves, 10 grams at 1.40 lire per kg. Onions, 25 grams at 0.60 per kg. Garlic, 25 grams at 0.80 per kg. Pepper, 1 gram at 3 lire per kg. Coffee, 15 grams at 3.50 lire per kg. Sugar, 22 grams at 1,20 lire per kg. Rum, 4 cl. at 3 lire per liter. Dried spices, 25 grams at 1 lire per kg.]

Such calculations illustrate the degree to which Italian colonials themselves were to be guided and managed by state technologies. The health of individual Italian bodies, conceived of as elements of the national population, thus becomes a primary site of deliberate state intervention. Contained within the colonies, the collective behaviors and habits of Franchetti's *vital subjects* become as predictable as average rates of rainfall in Asmara (347). Furthermore, like literary topoi, these "figurative regimes" reflecting the

crops up again in Franchetti's plea for colonial expansion in Asia Minor during World War I, "It's not necessary to recall here how important the phenomenon of emigration—in particular proletarian emigration—is in Italy. Until now, the hundreds of thousands of Italians who are emigrating today have no corner of earth where they may live under the shadow of the Italian flag" (Franchetti, "Prefazione" v).

meticulously regulated biological lives of Italian colonists constitute "the substitution of language for life." As Hunt and Rudolf remark, "life [...] becomes accessible to biopolitical intervention, not as such, but through its entry into language and representation." (20).

Whereas in Franchetti's southern writings, the ordering of sexual relations is aimed at a suppression of reproduction, within his colonial texts, Italian peasants are almost obsessively celebrated as vigorous, robust, and productive. Franchetti projects a "dense and immensely productive population" and a concomitant "intense production of wealth" upon Eritrean soil, forged by "that physical and moral vigor" that characterizes his ideal colonial laborer. He imagines that the Italian state will, "assure in few years to thousands of Italian peasant families economic well-being conquered with work, economic independence, and moral regeneration that they cannot obtain in Italy if not through laws that are unlikely to be approved, much less effective" (*Mezzogiorno e colonie* 391). Colonial conquest is an opportunity to enable the "healthy exuberance of [Italy's] energies and population" and to permit its "organic development" ("Prefazione," v).

Franchetti's colonial fantasy is formulated in part by the deployment of the topos of gendered African soil that lays itself open for penetration by Italian colonials: a revitalized Italian peasantry will become "economically, physically, and morally robust" because "the plain is offering its unoccupied lands to them" (*Mezzogiorno e colonie* 402). Each colonial family will be given, "virgin and fertile lands" to "sow" (321). If the Eritrean highlands are figured as a female eagerly awaiting Italian penetration, Italian colonial families themselves are to be "planted" in the soil (312–313).

What is most striking about Franchetti's feverish insistence on peasant productivity is precisely the paradox of biopolitics: in celebrating "the vital forces of a nation" Franchetti's texts are able to advocate unproblematically not only the state-subsidized redirection of Italy's landless, working poor, but, ultimately, the violence against and dispossession of Eritrean populations that his colonial project presupposes (411). In his seminal lectures on biopolitics, Foucault asks, "How will the power to kill and the function of murder operate in this technology of power, which takes life as both its object and its objective?" (*Society Must Be Defended* 254). His answer is racism. Foucault writes, "In the nineteenth century [...] war will be seen not only as a way of improving one's own race by eliminating the enemy race (in accordance with the themes of natural selection and the struggle for existence), but also a way of regenerating one's own race" (257). As Hunt and Rudolf describe Foucault's lectures, racism serves two functions: "first, the means to create biological 'caesuras within the [population]' that allow a line to be drawn 'between what must live and what must die' and, secondly, it creates a 'positive relationship' in which 'the death of the bad race ... is something that will make life in general healthier and purer'" (8). Indeed, anxieties about the regeneration of a population of Italians outside the juridical domain of the Italian nation-state emerge most explicitly within Franchetti's articulation of a colonial project:

> Ogni anno, abbandonano l'Italia, senza pensiero di ritorno, circa centomila emigranti, in massima parte contadini. Vanno con diversa fortuna a rinsanguare nazionalità straniere e, nelle presenti condizioni, non possono andare a fecondare col loro lavoro le terre fertili di clima mite che rimangono abbandonate sull'altipiano della nostra colonia. (*Mezzogiorno e colonie* 380)

> [Each year, without any thought of return, roughly one hundred thousand emigrants—primarily peasants—abandon Italy. They go with varied fortune to *revitalize foreign nationalities*. And yet, given the present situation, they are not allowed to *fecundate with their work the fertile and temperate lands that lay abandoned on the highlands of our colony*.] (emphasis mine)

Hundreds of thousands of Italian peasants embark on a one-way trip to foreign lands, infusing other national populations with new blood and labor-power (which, like the verb *rinsanguare* exists at the threshold of physiology and [political] economy), yet the fertile lands of the new colony are kept from being inseminated with Italian agricultural labor.[25] This passage marks one of the few deployments of an economy of blood within Franchetti's texts. "Rinsanguare" means literally "to supply or transfuse with new blood," and only figuratively to "revitalize," "boost," or "supply" economically. Significantly, this ambiguous term emerges when the southern question, emigration, and African colonization collide. Read symptomatically, this passage suggests that what is at stake in Franchetti's texts is indeed harnessing the productivity (in both senses) of an Italian race, lest

25 Franchetti returns to this formulation after the Italian invasion of Libya and before the conclusion of World War I, as he laments the fate of Italian emigrants inhabiting other colonial lands (particularly Tunisia): "In the remaining regions around [the Mediterranean Sea], occupied by other civilized nations, Italian activity is taking place, but our compatriots are destined to be absorbed sooner or later by the dominating nationality" ("Prefazione" iv).

this race begin to (re)produce in foreign lands as a result of emigration. Here, Franchetti espouses what Mark Choate characterizes as a widespread, popular "view of population control dating to the Middle Ages, [which understood] spontaneous emigration as a 'hemorrhage' of Italy's best blood, assuming that the most industrious of the poor would emigrate to better themselves." Choate continues, "In the Darwinian struggle between national peoples, the Italian race would falter unless the government retained its population, the basis for national survival" (*Emigrant Nation* 33).

For Franchetti, the triumphant regeneration of an Italian race can only happen "under the shadow of [the Italian] flag." As the Italian state has inhibited its landless peasantry from regenerating on its own soil, and since the Eritrean highlands seem to be "offering themselves" to laboring Italians, goes the logic, a state-sponsored redirection of these populations to colonial lands is the only solution. Yet, in order to execute a project centered upon the health, life, nutrition, and productivity of Italian bodies in colonial lands inhabited by local populations, the discourse must construct populations that potentially threaten the proliferation of this life. And it is at this point in the logic of the discourse that an explicit deployment of race becomes necessary:

> È fortuna che ci sia dato creare un tipo di società di razza italiana al di là dei mari, la quale abbia quel vigore che nasce, non tanto dalla lotta cogli elementi e con la natura vergine, quanto da uno stato di indipendenza economica [...] È questo il primo, il grande benefizio della nostra impresa africana. (*Mezzogiorno e colonie* 309)
>
> [It is fortunate that we have been given [an opportunity] to create a society of the Italian race overseas, the kind that possesses that vigor that is born, not so much from a battle with the elements or with virgin nature, as from a state of economic independence. [...] This is the first, the largest benefit of our African mission.]

The greatest outcome for the Italian colonization of Africa will thus be the production of explicitly racialized *vital subjects*. And what will be the main function of this "vigorous" Italian race—once a space for concentrated productivity has been established? Its own protection in a state of war: "We must send a part of the population that provides soldiers to the army and provides for their nutrition. Then we can truly say that the Italian Nation is in the colony" (392). The agricultural hyper-productivity that Franchetti has celebrated throughout his meditations on demographic colonialism and that is to enable settler families a viable economic income is thus recast as necessary to the reproduction and nutrition of healthy soldiers for the Italian colonial army. Furthermore, Franchetti's call for the proliferation of soldier-subjects underscores the degree to which the nation itself is produced by and through colonial expansion: the Italian nation finds its true expression in an African colony, and it exists only insofar as it is racially and agriculturally (re) productive. Indeed, just a few years earlier, Franchetti had extolled "military religion" as the highest "feeling for the *patria*," and had decried pacifism as a force that "weakened national energies" (Pezzino 66). Some years later, as the focus of Italy's colonial ambitions had shifted from Eritrea to Libya, Franchetti praised a 1912 law granting universal male suffrage as a way of allowing peasants, as the most essential "elements" of Italy's economic and military strength, to "feast at the banquet of the nation's growing prosperity" (Pezzino 64).[26] The *sentimento militare* (military feeling) was the highest form of patriotism, one that "[made] a people feel that they must be ready to give everything for the great ideal of the *patria*—even their own lives and those of their children" (Pezzino 65). The Italian state needed to complete "the work of unification, of fusion among all the forces that constitute the nation, and to these forces the union and the participation to the collective action of the country will give new life and new vigor for the increased grandeur of Italy" (Pezzino 64). Through sacrifice and death, the Italian national community was to find "new life and new vigor."

To be sure, Franchetti does not linger on reproductive policy or politics in crafting his rhetoric of colonial (re)productivity. Still, it seems safe to say that the "part of the population" that he claims must be "sent" to colonial Eritrea in order to reproduce and nourish healthy colonial soldiers are white Italian women.[27] In the earliest days of colonial conquest, as Franchetti was

26 "Sulla riforma della Legge elettorale politica. Discorso del senatore Leopoldo Franchetti pronunziato nella tornata del 26 giugno 1912" (Rome: Tipografia del Senato, 1912) 19. Cited in Pezzino 64–66.

27 Though Franchetti does not, I specify "white" here in order to begin to undo the normative fiction of Italian whiteness, and also to call attention to whiteness as a socio-political position. Quoting Ien Ang's 2003 article, "I'm a Feminist, but... 'Other' Women and Postnational Feminism," Derek Duncan uses the category "white" in his analysis of the colonial legacies of representations of Albanians in contemporary Italian film to refer not to biology or skin tone, but to a political position in a "structural, hierarchical inter-relationship" ("Italy's Postcolonial Cinema" 200). Furthermore, at this point in Italian colonial history, indigenous populations and colonial subjects were "by definition deprived of the rights of citizenship" (Barrera, "Sex, Citizenship, and the State" 158).

busy recording the daily alimentary intake of Italian colonial laborers and conducting agricultural experiments in the Eritrean highlands, there were very few European women in the colony.[28] Local Eritrean populations were of little to no interest to Franchetti in his fantasy of the colony as a vast, fertile land awaiting Italian labor and biological (re)productivity.[29] That Franchetti's project is one ultimately aimed at colonizing Italians themselves is thus made all the more clear by those reproductive bodies that his colonial reports omit but who were nevertheless a constant presence from the beginning of the Italian occupation of Eritrea—local women. And yet, in the early period of Italian occupation, sexual relationships in the form of concubinage or *madamismo* and prostitution between Italian soldiers and African women were quite widespread and encouraged as a means of maintaining the vitality or "wellbeing" of soldiers.[30] As early as 1885, one of the first orders of business in coordinating the Italian occupation of Massawa was regulating prostitution and opening a *sifilicomio*. Medical doctors and the Carabinieri were enlisted to identify a restricted pool of licensed prostitutes or "sanitized women" to be offered to Italian troops. While the sexual politics of colonialism would change with the arrival of Governor Ferdinando Martini in the wake of the defeat at Adwa in 1896, tending increasingly toward racial segregation, which would become more widespread after Mussolini's invasion of Ethiopia (1935–1936), in the so-called conquest years in which Franchetti was writing, official policy sought to encourage and regulate unions between Italian men and African women. Indeed, until the Ethiopian campaign, children born of these unions were granted citizenship if their Italian paternity was officially acknowledged. Clearly, these sexual politics of colonial rule, which would eventually be tethered to all kinds of notions about maintaining white "prestige" (and this well before fascism officially embraced biological racism in 1935), were of less interest to Franchetti than the imagined (re)productivity of Italians on which his Eritrean fantasy relied.[31]

28 Sòrgoni reports that in 1904, there were roughly 1,800 Italian men living in Eritrea, alongside 480 Italian women. By 1938, as a result of Mussolini's imperial project, there were roughly 67,000 "whites" (men and women) in the colony, as compared to roughly 596,000 locals, of whom roughly half were women. See Sòrgoni, *Parole e corpi* 29.

29 This might explain why, in her cogent discussion of liberal-era colonial anthropology and its interest in the sexual politics of colonial rule in Eritrea, Barbara Sòrgoni makes no mention of Franchetti.

30 On *madamismo* in East Africa, see: Ponzanesi; Iyob.

31 I have drawn my discussion of sexual policy in colonial Eritrea from Barrera, "Sex, Citizenship, and the State" and Sòrgoni, *Parole e corpi*.

Bringing the Colonies Home

Biopolitics, as a form of governmentality, entails an exercise of power as a kind of guiding of behaviors, or "the conduct of conduct." It draws its model for a political economy of forces from the private sphere of the home, or the *oikos* (as the etymological connection between *oeconomia*, household management, and modern political economy suggests). Modern institutions—schools, prisons, colonies, and the like—are often places where these two spheres intersect. As Suzanne Stewart-Steinberg has noted in her seminal discussion of Maria Montessori's social-maternal pedagogical method, the school, and debates around the nationalization of public instruction as a means of "making Italians," were particularly dense sites for the exercise of such power, particularly in the decades around the turn of the twentieth century. Montessori saw education as a social cure. "Montessori would later take a further step," writes Stewart-Steinberg, insofar as she would "bring the school into the home" (307) in her founding of Rome's famed Casa dei bambini beginning in 1907. The Casa grew out of a project of urban renewal in the San Lorenzo district, spearheaded by Edoardo Talamo and the Roman Association of Good Building, which turned slum buildings and shantytowns into "hygienic" proto-*borgate* or "new towns" just outside the Aurelian walls. According to Montessori, the project would "acquire city tenements, remodel them, put them into productive use, and administer them as a good father of a family would" (Montessori, cited in Stewart-Steinberg 320). At stake, Stewart-Steinberg argues, was a "national project of civilization, known as *incivilmento*," and Montessori's ideal teacher, as a "social mother," took on the role of socializing the family through "civiliz[ing] its members and open[ing] its walls to make its scientific management possible" (323). It should thus come as no surprise that, after his early inquiries into the southern question and his subsequent studies in the Horn of Africa, Leopoldo Franchetti dedicated the remainder of his life to the education of the peasantry as a means of making them into modern national subjects.

Alongside his wife, Alice Hallgarten, a wealthy American-born philanthropist, Franchetti founded the Scuola della Montesca in 1901 at his private villa in the Umbrian Città di Castello.[32] The school swiftly became an

32 The school operated continuously until the early 1980s, when it closed its doors due to under-enrollment. Today, the villa hosts a research center and a pedagogical institute, the Centro Studi e Formazione Villa Montesca, with a focus on technology in the classroom. In June 2014, the center hosted a screening of Paolo Malizia's historical

internationally renowned center for scientific and experimental pedagogy.[33] Franchetti saw the Montesca as an opportunity to continue the work he had been doing for nearly three decades toward the formation of a "populous and hard-working class of small landowning farmers" (Bonomi 10). The goal was, in Franchetti's patronizing words, to educate the "agricultural plebes" toward spiritual and social awareness, introducing them to a "new humanity."[34] In 1902, the couple opened a second school roughly ten kilometers from the first site, at Rovigliano. In August 1909, they hosted and financed the first international course on Montessori's scientific pedagogical method. Montessori had already been a recipient of Franchettian patronage as she wrote and published her groundbreaking treatise *Il metodo della pedagogia scientifica applicato all'educazione infantile nelle Case dei bambini* (The Montessori Method: Scientific Pedagogy as Applied to Child Education in the "Children's Houses") at their villa (and at their expense) earlier that year.[35]

Hallgarten, Franchetti, and Montessori had met in Rome, where Alice, a "practical feminist" of the early twentieth century, alongside other like-minded social interventionists such as Montessori, Linda Malnati, Aurelia Josz, and Felicitas Buchner, was an active member of the Unione per il bene (Union for Good), also in Rome's impoverished San Lorenzo neighborhood. Hallgarten Franchetti was an organizer interested in hands-on philanthropy, and took a special interest in hygienic instruction and care for women and children. Born to a bourgeois family of bankers, life at the Montesca introduced her to the joys of manual labor: "I would have never believed that an industry [...]

docu-fiction film *Alice Hallgarten e Leopoldo Franchetti. Una visione condivisa.* See: http://www.montesca.it/index.asp (accessed July 9, 2014).

33 Similar centers were emerging throughout Italy in these years, such as the Scuola materna di Rosa e Carolina Agazzi; Colonie dei Giovani Lavoratori (David Levi-Morenos); Scuola inventive (Gino Ferretti); Scuola di educazione dell'attività spontanea (Maurilio Salvoni). Sante Bucci argues that the Scuola della Montesca was, rather than a regional center, part of an international network associated with the New Education movement founded by John Dewey. See Bucci. For more on the intellectual context of Hallgarten's pedagogy, see Waldbaum.

34 Enrico Zangarelli, *Leopoldo e Alice Franchetti. La scuola della Montesca,* Prhomos, 1984: 80 (cited in Moretti 141).

35 Montessori dedicated the original Italian volume *Il metodo della pedagogia scientifica applicato all'educazione infantile nelle Case dei bambini* (Città di Castello: S. Lapi, 1909) to the Baron and Baroness Franchetti and later the 1912 English edition of her magnum opus, which Franchetti also financed, to the memory of her then recently deceased friend and colleague Alice Hallgarten (Waldbaum 129). Hallgarten died of tuberculosis at the age of thirty-seven.

could be so pleasant," she wrote. "Manual labor is good for the soul and forges a bond of understanding among all" (Fossati 291). Hallgarten Franchetti was a "social mother" par excellence, as she and the Baron Franchetti never had children of their own, and on her deathbed she penned letters asking after "i (suoi) bambini amati" ("(her) beloved children") (Waldbaum 132–133).

If, as Stewart-Steinberg notes, the Virgin Mary was the model for a secular, social Maria in what quickly became the cult of Montessori, Hallgarten Franchetti was the "fiammella francescana" ("little Franciscan flame"), as her friend, colleague, and early biographer Aurelia Josz affectionately named her. Montessori herself referred to Alice as a "saint" (Bucci 202). Alice shared with her husband a common interest in botany, agricultural experimentation, and moral and social regeneration through a return to rural life, which was inspired by her studies of John Ruskin, Leo Tolstoy, and St. Francis of Assisi. Hallgarten Franchetti herself penned a daily prayer to open each elementary school day modeled on St Francis's thirteenth-century *Canticle of the Sun* (Waldbaum 130). For her primary contemporary biographer, Roberta Fossati, Hallgarten Franchetti's social utopia, which linked her to many other women philanthropists and social advocates of her day, included "the loving cultivation of the earth, divided up into small plots to be tended personally, the moderate use of brick for the construction of houses" (194), and weaving and spinning with techniques allegedly inherited from Giotto and Homer.

The Montesca was home to an elementary school that served the local peasantry around Città di Castello, most of whom labored on the Franchetti's land (Waldbaum 130). Erica Moretti has highlighted the ambivalence of the Franchetti's pedagogical experiment, arguing that in spite of their avowed focus on the education of peasant children as a means of social emancipation, course assignments and lesson plans reveal an ideological project aimed at tethering sharecroppers' children indefinitely to the land (Moretti 144–145). See Fig. 1.1.

The Franchetti's fostered pride and passion for the agricultural work of the students' forebears through dictation exercises that emphasized manual labor as the "inevitable law of all." Student notebooks from the school's first decade contain multiple dictations on the theme of "Work" that extoll the virtues of hard work across the animal kingdom, or refer to labor as "a duty and a right" while condemning "loafers who live on the money of others" as "unworthy of being called men." Workers, the Franchettis' students diligently transcribed, must take care not to "squander their physical and intellectual energies [*forze*] too early in life," lest they become a burden to the society, "in which and for which they live." In addition to labor, in-class writing assignments focused upon public hygiene and domesticity (Fig. 1.2), with

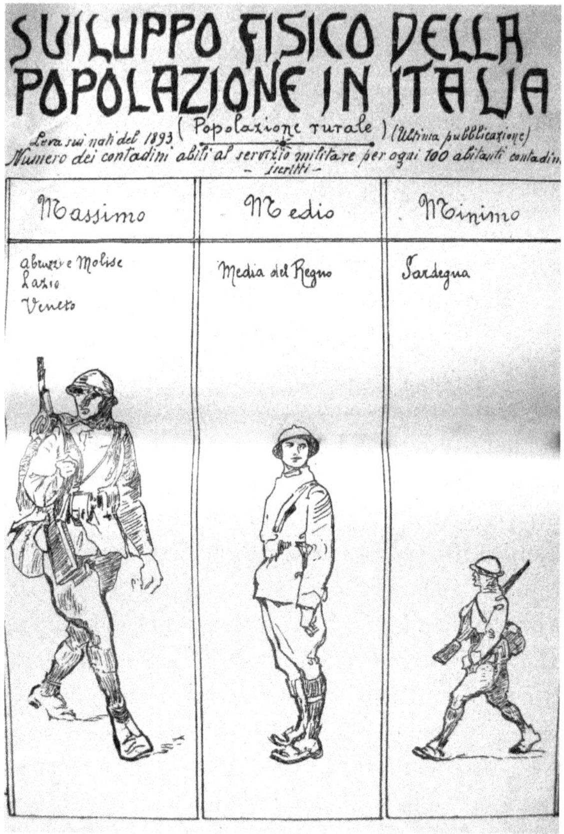

Fig. 1.1 Among the didactic materials from the Montesca and Rovigliano schools: a sketch illustrating "Physical Development of the Rural Italian Population, based on those born in 1893. Number of peasants fit for military service" (Photo credit: Erica Moretti.)

essays such as "How I Would Make a Home," which describe a three-story house flooded with natural light, on arable land, with room for a cellar and stables for oxen.[36]

Although the school at Villa la Montesca began as an elementary school, it soon expanded to include training for male laborers in agricultural techniques,

36 The Archivio unico di deposito della Regione Umbria contains primary sources such as textbooks, student notebooks, and other pedagogical materials used at the Schools of Villa Montesca and Rovigliano. I am incredibly grateful to Erica Moretti for sharing these resources with me.

Fig. 1.2 Schoolbook on "The Benefits of Hygiene" (Rovigliano School, 1909) (Photo credit: Erica Moretti.)

and courses for women in weaving, "domestic economy," and hygiene. In an era of increasing industrialization, the Franchetti's pedagogical project to some degree pushed back, romanticizing rural labor, as well as its correlate, artisanal manufacturing. Beginning in 1908, unemployed or otherwise impoverished women would come to train in Hallgarten Franchetti's Laboratorio Tela Umbra weaving facility, and would bring their nursing infants and children to the adjacent nursery and elementary schools.[37] During winter breaks, peasants would receive instruction in woodworking in order to help them furnish their

37 The Tela Umbra operates to this day as an artisanal weaving studio. The palazzo also hosts the Museo Franchetti, dedicated to its founders. See: http://www.telaumbra.it/ (accessed July 9, 2014).

homes. Women were to learn "sapere pratico" (practical knowledge), which would improve the management of the household in daily life or provide professional development in the event women were to seek work as *governanti* (governesses) (Fossati, "Il lavoro culturale" 292–294). This method of social intervention, much like Montessori's and that of Paolo Mantegazza (the subject of Chapter Two), brought the government of the household in line with that of the new national community through the deliberate production of certain kinds of knowledge, or *savoir.* Franchetti's writings on the southern question and early colonization in Eritrea are also laden with such a biopolitics of knowledge. Indeed, what prompted his wife to attend to the training (read: governing) of peasant and impoverished women was the failure of another of Franchetti's attempts at colonization: a short-lived Colonia Agricola on Rome's via Flaminia (also known as the Dormitorio Sonzogno), which provided agricultural training to orphans and delinquent youths and opened in 1899.[38] When the Roman colony was closed two years later, Franchetti transferred the urban youths to another of his properties, the Buon Ricovero estate, and unsuccessfully tried to instill in them a love for rural labor.

Franchetti would have been familiar with the pioneering work of the Catholic priest Leonardo Murialdo (1828–1900) on agricultural colonies as transitional but ultimately carceral programs for young offenders. Founder of the vocational school Pia Società Torinese di San Giuseppe, Murialdo had spent time in England, Belgium, and France visiting agricultural orphanages, or *orphelinats agricoles.* The first Italian agricultural colony opened at Moncucco Torinese in 1853 (Bucci 214).[39] By 1872, there were also agricultural colonies operating in Assisi, Bosco Maregno (Piedmont), Perugia, and Scansano (Tuscany). Critically, while young male offenders were confined in agricultural colonies, young women were relegated to the domestic sphere; theirs were rehabilitative homes, or *case*: the Pia Casa di Nazaret in Milan; the Casa di Patronato in Turin; the Casa di Riabilitazione in Venice, for example (Ministero dell'Interno 430). Leone Carpi's 1874 *Delle colonie e dell'emigrazione d'italiani all'estero* (On the Colonies and Emigration of Italians Abroad), one of Franchetti's source texts for emigration figures in *La Sicilia nel 1876*, calls for the establishment of a system of penal agricultural

38 Franchetti was the president of the Colonia Agricola. Don Brizio Casciola, a protestant literary figure and social interventionist who had greatly influenced the Franchetti's (both of whom were secular Jews by birth) was the colony's director.

39 Italy's last agricultural penal colony still operates on the island of Gorgona, off the coast of Tuscany (Alessandrini).

colonies in Italy, with the goal of moral and political rehabilitation and education, which he argues would be consistent with "modern civilization [*civiltà*]." He outlines the targets of such correctional institutions: 1) at-risk orphans and foundlings; 2) corrupt or convicted minors and adults convicted of minor offenses; 3) "gli accattoni [...], gli sfacendati, i girovaghi ed i viziosi" (scroungers, loafers, vagrants, and perverts) (Carpi 140). Franchetti's turn to agricultural penal and pedagogical colonies did not so much signal an abandonment of his colonial project in the Horn of Africa, as much as the coexistence of these various modes of agricultural colonization. Indeed, as the Franchetti's school at Montesca was continuing to expand its reach among the Umbrian peasantry, his African colonial fantasy endured.

Caffé, Cafoni, Cacao e Cotone: Turning Desperation into Design

In July 1907, Doctor Gino Bartolommei Gioli, along with Franchetti, inaugurated the publication of a new monthly periodical entitled *L'Agricoltura coloniale* by outlining its mission.[40] As the official organ of the Institute for

40 The editorial history of *L'Agricoltura coloniale* in many ways reflects the ideological and rhetorical trajectory of colonialism in Italy (as elsewhere). From its inception through to the end of World War II, *L'Agricoltura coloniale* enjoyed an explicitly colonial gaze, published by the Istituto Agricolo Coloniale Italiano. In January 1945, its title was changed to *Rivista di agricoltura tropicale e subtropicale,* reflecting the imposition of a more nebulous ideological frame upon the symbolic geography of Africa: colonial space, characterized by inherently hierarchical power relations, becomes climatic space, where atmospheric conditions, rather than armed men, reign. In 1953, the publishing Institute changed names, to the Istituto Agronomico per l'Oltremare (a name which remains today). In January 1998, the publication underwent another transition, as it became an English-language journal entitled *Journal of Agriculture and Environment for International Development.* The symbolic trajectory of Africa, from a space of colonial desire in the nineteenth and early twentieth centuries to one of economic aid and capitalist development in the late twentieth and twenty-first, is legible not only within this reductive history of an institution and its monthly periodical, but within a variety of representational practices, from the naming and renaming of formerly colonial institutions and offices to the appropriation of the colonial topos of *mal d'Africa* by contemporary Italian human aid groups working in Africa and the ad campaigns of Benetton. These expressions of the afterlife of colonialism are not proof that desire has been purged, but instead illustrate the degree to which it has been reworked to accommodate, rather than fantasies of domination through violation, something closer to a no-less colonialist hegemony, as nations and peoples are figured as either capable of offering aid or desperate to receive it.

Italian Colonial Agriculture and the Office for Experimental Agriculture in Eritrea, the primary goal of the journal was to ensure the "technical instruction of future agrarian colonizers" (Gioli 2).[41] The introductory address by Gioli and subsequent articles by the bourgeois professional class—senators, professors, scientists, and doctors—which fill the pages of *L'Agricoltura coloniale* illustrate the breadth in the liberal period of the concept of "colony," which signified, variously: 1) the "spontaneous" colonies of farmworker emigrants in the Americas; 2) colonies of "direct dominion" in East Africa; 3) agricultural, educational, and/or penal colonies within Italy. What drew these fields together was agricultural labor. For Gioli, Franchetti, and their colleagues, the outpouring of Italian labor-power to the Americas needed to be harnessed and managed through instruction. As such, the journal also publicized a new degree program, which would train professionals in colonial agriculture, botany, economics, geography and history, the hygiene of humans and of livestock, emergency medical care, and foreign languages.

At stake was Italy's position with respect to the established colonial powers Britain, Holland, France, and Germany. Up to this point, emigration and the "spontaneous" colonization of the Americas had been the work of poor, landless, and uncultured Italians, whose exploited status in the Americas risked resembling that of the "negro slaves of old" (75). Italy had to prove that it was capable of producing not only *cafoni* (poor, landless peasants) but also a class of managers competent enough to direct their labor. Hygiene was a constitutive part of this program, as colonists were conceived of as physiological beings who faced a variety of specific and organic challenges due to the changes in climate and soil inherent to relocation. As Doctor Enrico Persano put it in his article, "Colonial Hygiene":

> The colonial problem, reduced to its most simple expression, may be posed in these terms: white men, born and raised in temperate regions, and used therefore to living within the climatic and telluric conditions of such regions must, as quickly as possible, get used to living, prospering, and producing in tropical or even equatorial regions, under completely different climatic and telluric conditions, which [...] have often proved to be fatal for them. (123)

41 Unless otherwise noted, this and all subsequent translations from the Italian are my own.

Yet, if the hygienic discourse presented in *L'Agricoltura coloniale* affirms a certain degree of physiological unity (Italian colonials are unequivocally "white men"), this wholeness is proclaimed through a rhetorical dismemberment: able-bodied worker-colonists are figured overwhelmingly as little more than "arms." The movement of laboring bodies is referred to as "the emigration of arms" (Gioli 4).[42] As in Prime Minister Crispi's language in the Introduction ("colonies must be like arms"), the emigrant peasant body exists only through its relationship to labor productivity. Furthermore, the emigrant population is fetishistically made whole through its rhetorical truncation. The rhetoric of loss functions fetishistically to 'mend' the constitutive absence at the heart of the racialized Italian national subject. Truncated limbs will recur again in D'Annunzio's occupation of Fiume, though they will be recast as belonging to wounded soldier-patriots whose dismemberment, like that of Italy's after World War I, is "healed" by aerial conquest.

By 1890, when Prime Minister Crispi led efforts to secure the colony, there had been a shift in the rhetorical and ideological registers of Italian colonialism in the Horn of Africa: while earlier discussions had centered around colonization for diplomatic and/or military gain, the Eritrean venture marked the advent of a new official disposition (Rainero, *I primi tentativi* 11).[43] Within parliamentary discussions of "demographic colonialism," agriculture came into focus as both the target and the apparatus of Italian rule. As the inaugural issue of *L'Agricoltura coloniale* illustrates, agriculture as a *dispositif* enabled the cohesion of a set of terms under a new sign: that of biopolitics.[44] Peasant bodies were conceived of not only as potentially productive agricultural laborers (Marx's labor-power), but also as elements of a biological Italian population, whose reproductive capacities and alimentary intake were

42 Lauding Italy's Libyan campaign in 1911, poet Giovanni Pascoli would refer to Italy somewhat melancholically as "the great provider of cut-rate arms," referring to the mass emigration of laboring Italians. See: Chapters Three and Four of this study.

43 Rainero also offers an account of the failure of Franchetti's project. For a similar appraisal of the outcome of Franchetti's plan, as well as a detailed description of it (in English, based on Rainero's account), see Larebo 12–19.

44 I draw here from Foucault's *dispositif* (typically translated as "apparatus"). The *dispositif* consists of the relations between "a thoroughly heterogeneous ensemble consisting of discourses, institutions, architectural forms, regulatory decisions, laws, administrative measures, scientific statements, philosophical, moral and philanthropic propositions" (Foucault, *Power/Knowledge* 194–195). See also: Agamben, *What is an Apparatus?*

sites of discussion, projection, and textual production among policymakers. "Demographic" colonialism, a uniquely Italian contribution to European colonial thought and practice, therefore refers not simply to the Italian settlement of colonies in East Africa, but to a reproductive project aimed at Italian settlers as *vital subjects*. To refer to Franchetti's colonial project is thus to draw together the three overlapping biopolitical fields of social intervention to which he dedicated his life-work: the southern question, "demographic" colonialism in the Horn of Africa, and a pedagogical model of guiding, or instruction, that culminated in the "corrective" (disciplinary) agricultural colony.

From Anti-racism to Biopolitics

Franchetti's writing and social intervention illustrate how state practices centered around what David Horn identifies as "the emergence [in the 1920s and 1930s] of new social technologies – including censuses, social insurance, practices of urban planning, housing projects, and social work" (*Social Bodies* 4) were rehearsed in southern Italy and the colonies (both abroad and domestic) well before the fascist state practiced them at home. What is particularly striking is that, contrary to Horn's claim that the rise of such technologies resulted from the fascist state's concern with declining fertility, the emergence of the problematic of the population was instead already in progress during a time of fervent biological productivity and that this problematic was informed by a logic of race war long before the fascist regime's official implementation of the racial laws in the metropole. Indeed, in 1937, Italy's first racial laws declared sexual relations between Italian citizens and colonial subjects illegal and punishable by five years of imprisonment (Barrera, "Sex, Citizenship, and the State" 157). Franchetti's texts thus illustrate one of the many routes of what Foucault called the *effet de retour* of European colonialism on models of power in the metropole. In one of the few moments in his seminal 1976 lectures on biopolitics in which he mentions colonialism, Foucault states:

> It should never be forgotten that while colonization, with its techniques and its political and juridical weapons, obviously transported European models to other continents, it also had a considerable boomerang effect [*effet de retour*] on the mechanisms of power in the West, and on the apparatuses, institutions, and techniques of power. A whole series of

> colonial models was brought back to the West, and the result was that the West could practice something resembling colonization, or an internal colonialism, on itself. (*Society Must Be Defended* 103)[45]

If it is reasonable to assume that Foucault did not have Italian colonialism in mind here, Franchetti's writings nevertheless constitute an example of this dynamic "internal colonialism," given how he writes about Italy's southern regions and how he approaches the ultimately self-referential demographic colonialism in Eritrea, and later the agricultural penal colony in Rome and Buon Ricovero, as well as the training facility at the Montesca. Franchetti's writings illustrate not only a "boomerang effect," but a constant shuttling back and forth between colonizing Italians both within and outside the ever-shifting borders of the nation-state (in terms of both knowledge production about Italy's southern regions and pedagogy). At its most basic, his colonial model points to the profound imbrication between colonial discourse and the project of "making Italians" through the biopolitical rhetorics of race and (re)productivity. Italian colonialism—understood not simply as African conquest, but as a wider project of colonizing Italians themselves, through training or Foucauldian "guiding" in domestic hygiene, alimentation, and agricultural labor—was viewed by Franchetti and others as a means of healing national fragmentation through a harnessing of Italian (re)productivity.

At least since Alfredo Niceforo's 1898 declaration that the "barbarous" Italian South was "a great colony to civilize" (6), and certainly since Antonio Gramsci's 1927 polemic against what he described as the colonial subjugation of the South and the islands to the Italian mainland, scholars of Italy's southern question have acknowledged the role that colonial modes of representation have played in its historical trajectory (*La questione meridionale* 132).[46] In spite of this fact, recent studies by Nelson Moe and John Dickie have focused exclusively upon Franchetti's southern texts, without taking into account his concomitant interests in "demographic" colonization in Africa and domestic pedagogy (Moe, *The View from Vesuvius*; Dickie, *Darkest Italy*). Furthermore,

45 The French *effet de retour* does not appear in the original English translation that I have been citing. I borrowed it from Stoler, who cites this passage in an English translation of the Italian. In her version, the "boomerang effect" is translated as "return effect" and the French *effet de retour* is given in parenthesis (Stoler, *Race and the Education of Desire* 75).

46 For an analysis of Gramsci's contribution to the southern question in its historical and political context(s), see Davis, "The South, the Risorgimento, and the Origins of the 'Southern Problem'". See also Verdicchio, "Introduction."

when historians such as Romain Rainero, Angelo Del Boca, and Nicola Labanca have addressed Franchetti's colonial plans, they tend to be depicted as beneficently liberal, nationalist, and/or "demographic," and thus as not-yet racist, not-yet imperialist, and, therefore, not-yet fascist.[47] As Dickie puts it with regard to Franchetti's southern writings, "[Pasquale] Villari and Franchetti are often viewed, from both left and right, as the origin of a long and honourable tradition of social analysis anchored in objectivity" (*Darkest Italy* 55). Similarly, scholars of Italy's southern question (e.g. Salvadori 184) tend to split the corpus in two, cordoning off what are figured as the liberal-humanitarian works of Pasquale Villari, Leopoldo Franchetti and Sidney Sonnino, and Giustino Fortunato from the explicitly racialist writings of Alfredo Niceforo, Giuseppe Sergi, and Cesare Lombroso. The more or less blatant racism of both colonialist and meridionalist thinkers thus becomes the primary index of how they are remembered by historiography and, consequently, by their modern readers. And yet, the reading of Franchetti conducted in this chapter problematizes some of these dearly held assumptions about Franchetti's place in the genealogy of Italian race thinking. As the coming chapter illustrates, the anti-fascism that propelled Italian anti-racism, while politically and even ethically necessary, has produced its own warped lens. If readers "know" in advance that Franchetti was neither a proto-fascist nor a racist—or at least not as much of one as Niceforo, Sergi, or Lombroso—they are inclined to perform a selective reading, overlooking some details of his text, while emphasizing others. One of the goals of this chapter has been to explore the degree to which Franchetti's plan (supported as it was, at least initially, by Prime Minister Francesco Crispi and others) provided the rhetorical and ideological conditions of possibility for Mussolini's warning to prefects on August 10, 1927 (Fig. 1.3):

> Per ogni emigrante che esce per sempre dall'Italia, in compenso di poco oro che giunge dall'estero, il paese perde economicamente tutto ciò che ha speso per nutrirlo, per educarlo, per metterlo in grado di produrre.

47 See, for example, Massimo Salvadori, who writes: "Despite his time in Parliament, Franchetti always remained a humanitarian. And we might say that [...] Franchetti experienced the political moment by following humanitarianism, the only truly deep feeling he had." For Salvadori, Franchetti's colonial project was the logical outcome of this humanitarianism: "It isn't too difficult to understand how [...] he and Sonnino boarded the train of imperialism. [...] Franchetti brought his humanitarian spirit to the question and saw in the colonies above all a way out of Italian poverty" (109).

Fig. 1.3 Mussolini during the fascist occupation of Ethiopia (1935) on the cover of the journal *L'agricoltura coloniale*, founded by Franchetti and Gioli in 1907 (Photo by author.)

Militariamente un soldato, demograficamente un elemento giovane e forte, che feconderà terre straniere e darà figli a paesi stranieri. (350)

[For every emigrant that leaves Italy permanently, in exchange for the bit of gold that arrives from abroad, the country loses economically everything that it spent to feed him, to educate him, to make him capable of producing. Militarily a soldier, demographically a young and strong element, that will fecundate foreign lands and give children to foreign countries.][48]

48 As David Horn points out, "[In] a variety of ways, Fascism took up the language of the medical and social sciences in a self-conscious effort to constitute itself as a modern form of government. However, it would be unwarranted to characterize as fascist the new discourses and practices that took the Italian population as their object. They were

Read symptomatically, and therefore in spite of their dry rationalism and apparent lack of rhetoric, Franchetti's liberal-humanistic texts figure the relationship between the potential (re)productivity of individual bodies (conceived of as elements of the Italian population) and emigration in precisely the same way as a justification for his early colonial project in Eritrea. How is it that the very same biopolitical rhetoric can be used to confirm the implicit racism of Mussolini's demographic policy, on the one hand, and the rationalism of Franchetti's liberal-democratic humanitarianism on the other? This is just one of the analytical pitfalls of Italian anti-racism that I propose we begin to undo.

part of a modern rethinking of society and the social in the late-nineteenth and early-twentieth centuries that tended to cut across both national and political boundaries" (*Social Bodies* 8).

CHAPTER TWO

Immunitary Technologies

> Paolo and Maria saw other fractures and dislocations, which were medicated painlessly and with the greatest ease.
>
> Paolo Mantegazza, *L'Anno 3000. Sogno* (141)

Leopoldo Franchetti's brand of colonialism envisioned agricultural productivity as the remedy to the biopolitical fragmentation of Italians. Fertile land and the nourishment colonial bodies and would-be soldiers stood to extract from it functioned as prophylaxis—defending the *razza italiana* both from further mutilation (departure for foreign lands) and from its decimation by local opposition. While the biopolitical rhetoric of defense may be almost self-evident in the colonial context, what happens to it when the potential threat comes not from outside, but from within? For Italy's preeminent Darwinian physician, public hygienist, and anthropologist Paolo Mantegazza (1831–1910), the Italian body politic was sick. Pathological bodies produced pathological politics, and ensured Italy's inferiority on the European, and global, stage. His answer to preventing the spread of a generalized infection was to write—hundreds of volumes, popular manuals, and pamphlets that would educate Italians about correcting and maintaining the proper function of their bodies. If the avowed goal of Mantegazza's popular scientific production was the protection of life, his final novel *L'Anno 3000. Sogno* (The Year 3000. A Dream), published in 1897, illustrates how that protection becomes its negation through the futuristic invention of immunitary technologies that "eliminate" pathological newborns in order to strengthen the remainder of the population.

Fig. 2.1 Entrance to the University of Florence's Museum of Anthropology and Ethnology, founded by Mantegazza in 1869 (Photo by author.)

Writing as Prophylaxis

In 1868, as Italian troops continued their struggle to establish Italy as a territorial whole, Paolo Mantegazza published a best-selling epistolary romance novel that inaugurated his ascent to literary celebrity. *Un giorno a Madera. Una pagina dell'igiene d'amore* (A Day in Madeira: A Page from the Hygiene of Love) traces the melancholy and ultimately unreproductive love affair of its protagonists, Emma and William, and can be situated within emergent discourses of national public hygiene, which were shaped in large part by Mantegazza and his colleagues at the Universities of Pavia and Florence.[1] By 1899, there were more than twenty editions of *Madera*

1 Mantegazza was indefatigable not only as a writer, but also as an academic and an elected official. From 1860 until 1869, Mantegazza was a professor of General Pathology at the University of Pavia. In 1865, he was elected parliamentary deputy of Monza. In

in circulation (Pasini 249–250).[2] Through the epistolary exchange that constitutes the bulk of the novel, readers learn that frail Emma is the last survivor in a family ravaged by tuberculosis.[3] On his deathbed, her father curses his "poisoned blood" and insists that Emma vow never to take a husband, much less reproduce, lest she perpetuate the diseased family line. The novel ends with William and Emma's relationship unconsummated and Emma's death. Her vow of physiological sterility thus upheld, she implores William in her final letter to "render [his] life fecund with courageous and great works" (*Un giorno a Madera* 182).[4]

Emma dies a heroine because—fated to die from consumption—she courageously refuses to reproduce. Her plight is paralleled by the threat of Italian consumption with which Mantegazza prefaces the novel. In his dedicatory notes to the Monza electorate, Mantegazza expresses anxiety about his role in Italy's nascent parliamentary administration:

> Poveri noi, se i nostri figliuoli dovessero trovare che l'opera dei primi deputati del Regno d'Italia andò tutta consumata nel fare delle mozioni sospensive, degli ordini del giorno puri e semplici e delle questioni pregiudiziali. Poveretti noi, se tutta la vita d'una generazione dovesse

1869, he moved to the University of Florence where he occupied Italy's first chair in Anthropology. Historian Giovanni Landucci identifies Mantegazza's role as central to a Florentine milieu engaged in intense debate surrounding anthropology and public hygiene. Furthermore, Walter Pasini notes that Mantegazza claimed to have written his wildly successful *Un giorno a Madera* in a matter of hours during a break in a parliamentary session (229). In addition to reflecting Mantegazza's recurrent rhetoric of effortless authorial prowess, this claim also illustrates how his novel must be read while bearing in mind his direct participation in shaping state practices of public hygiene.

2 From 1910 until 1924, *Un giorno a Madera* continued to be reprinted annually, in several cases by numerous publishers (Bemporad, Bernadoni, Brigola, and Treves were among his most faithful). New editions continued to appear, though at a far less steady rate, from 1926 to 1952. The most recent edition was published in 1991 by Edizioni ECIG (Genoa). A portion of the novel was translated into English by David Jacobson and published in Pireddu, *Paolo Mantegazza. The Physiology of Love and Other Writings* 351–376. Whenever possible, I have used Jacobson's translation.

3 The etiology of tuberculosis was not discovered until 1882 (by Robert Koch). Until then, tuberculosis was a major impetus for projects of public hygiene in Italy. Koch also made significant contributions to scientific understandings of cholera (W. Pasini 230). For two rich accounts of the role that public health has played in shaping social, political, and scientific history in Italy, see: Snowden, *The Conquest of Malaria; Naples in the Time of Cholera.*

4 Unless otherwise noted, translations are my own.

> andar consunta nel rattoppare i nostri cenci, nel puntellare le casse dell'erario, nel lasciare ai futuri della carta e dei debiti. Ognuno di noi deve aprire un solco in quella terra in cui i figli hanno a seminare il pane dell'avvenire. Questa terra bagnata di sangue l'abbiamo a fecondare del nostro sudore; e chi ebbe dagli elettori la più alta missione che si possa affidare a un cittadino, ha maggiori doveri degli altri di preparare la terra per una Italia migliore. (vi)
>
> [Poor us, if our sons were to find that the work of the first deputies of the Kingdom of Italy was completely *consumed* by putting things off, by carrying out the plain and simple orders of the day [...]. Poor little us, *if the entire life of a generation were to be consumed patching our rags*, shoring up the chests of the treasury, leaving bills and debts to our children. *Each of us must plow the land in which our sons may sow the wheat of the future. We must fertilize this blood-soaked land with our sweat*; and he who is granted the highest mission entrusted to a citizen by the electorate has even more of an obligation than others in *preparing the land* for an improved Italy.] (emphasis mine)

By deploying the metaphor of agricultural labor to describe newly unified Italy's need for sound policymaking ("each of us must plow the land in which our sons may sow the wheat of the future"; elected officials must "prepare the land for an improved Italy"), Mantegazza dresses his parliamentary deputies, heroic veterans and noble voters in peasant's rags (*cenci*) in order to figure his productive Italian population. He mobilizes this rhetoric of agricultural productivity in order to advocate a project of moral-physiological regeneration.[5] Consumption threatens the unproductive. A critical link is thus established between terrain—once bathed in bellicose blood, now to be inseminated with the figurative sweat of intellectual and political labor—and prospective genealogy, which he evokes by singling out "our sons [...] of the future." Here, as in Franchetti's colonial project in Eritrea and in D'Annunzio's Fiuman discourses, one finds the ineluctable infusion of the nationalized soil with the biological traces of its inhabitants. In each of these cases, labor, be it agricultural or intellectual, saturates and legitimates territories, fusing them

5 Throughout his oeuvre, one of Mantegazza's primary concerns is positing a relationship between morality and physiology. This project is best illustrated in Mantegazza and Neera. His physiology texts *Fisiologia dell'amore* (1873) and *Fisiologia della donna* (1893) are extended treatises on morality.

organically with the bodies that inhabit them. Mantegazza's patriotic trajectory is ultimately one of war, work, rebirth (or "resurgence"), and reproduction:

> E quando dico un'Italia migliore, voglio dire degli Italiani più sani e più onesti prima di tutto, poi più operosi e più sapienti, che è quanto dire più ricchi e più potenti. (vi)
>
> [And when I say a better Italy, I mean Italians who are more healthy and honest first and foremost, and also more hard-working and knowledgeable, which means more rich and more powerful.]

He clarifies his metonymic substitution ("Italy" for "Italians") in order to isolate his focus on the physiological and moral health of the population.[6] True to biopolitical form, Mantegazza meanders, though not without purpose, from biological fitness to political strength. This increase in the physiological strength of Italians ("more healthy," "more hard-working") presumably wards off potential encroachment from elements residing outside the national borders—the threat alluded to with the evocation, cited above, of Italy's not-so-distant past as a "land wet with blood." His rhetoric of productivity thus cedes to one of biological defense.[7]

Mantegazza's address to voters offers a figure for what Giorgio Agamben calls an "original biopolitical fracture": peasants—representatives par excellence of liberal Italy's rhetoric of the "people" ("the wretched, the oppressed, and the vanquished")—provide a rhetorical field in which proper citizens may be represented (Agamben, *Means without End* 30: 1). Politically disenfranchised peasants, as rhetorical placeholders, are thereby included in

6 The unveiling of his substitution of "Italy" for "the Italians" merely exposes another order of substitution: "the Italians" stand in for the desire of a body politic that has overcome its constitutive fracture.

7 Giorgio Agamben departs from Foucault's formulation of biopower, identifying instead what he calls a "fundamental biopolitical split" within the structures of Western political theory. For Agamben, this rift—between the "People" as the phantasmagorically integral body politic and the "people" as the marginalized, who threaten this unity from within—is precisely what Nazi Germany sought to remedy through its purging of the latter: "Nazism tried obscurely and in vain to free the Western political stage from this intolerable shadow so as to produce finally the German *Volk* as the people that has been able to heal the original biopolitical fracture." Agamben's "people" represent, "that naked life that modernity necessarily creates within itself but whose presence it is no longer able to tolerate in any way" (*Means without End* 33: 4).

a formulation that simultaneously marks their exclusion. The work of Italian regeneration, of staving off the consumption of intellectual and physiological force, is entrusted not to these (non-)subjects, but instead to servants of the state. Left in the wrong hands, the passage warns, attempts at regeneration risk exhausting Italian potential. In *Un giorno a Madera,* this attempt is figured as the threat of reproducing the biologically unfit, and thus blocks the love affair and affirms Emma's role as a respectable heroine.

Mantegazza grafts Emma and William's respective struggles between duty and desire onto racialized national difference. William is both doubled and halved ("a doubled man," half English and half Italian): "In me, love welded together two races [*razze*], two destinies, two worlds. [...] In every move, I feel in me Vesuvius and the London fog" (*Un giorno a Madera* 46).[8] While desire is figured as Italian, duty is English: "I feel like an Italian, I take action as an Englishman does" (Pireddu 357). When William pleads with Emma to ignore the orders of her dead father, she implores him: "Extinguish Vesuvius, William, and become English once again" (*Un giorno a Madera* 46). Presumably, were the "Italian" to triumph over the "English," the relationship would be consummated, resulting in the generation of a biologically unfit being. Mantegazza's novel, considered alongside his voluminous writings on national physiological and moral hygiene, may be read as a cautionary tale: if Italians reproduce before they are rendered, "more healthy and honest first and foremost, [...] more hard-working and knowledgeable, [...] more rich and more powerful," they will generate physiological and political pathology. The tuberculoid baby that haunts *Un giorno a Madera* is thus a figure for a degenerate Italy.[9] Yet Mantegazza's novelistic meditations on the biopolitical did not end here. In 1897, the threat of the pathological infant resurfaced, as Mantegazza arguably picked up where he had left off nearly thirty years earlier, publishing his futuristic eugenic utopia *L'Anno 3000. Sogno.*[10]

8 Jacobson's translation reads, "I feel two natures within me, two worlds of thoughts, sensations, joys, and sorrows. At every stroke I feel Vesuvius and the London fog" (Pireddu 357).

9 Both David Horn and Suzanne Stewart-Steinberg's recent readings of Cesare Lombroso describe his criminological project as shaped by similar anxieties about Italy's relationship to modernity. See: Horn, *The Criminal Body*; Stewart-Steinberg.

10 In her introduction to the English translation, Nicoletta Pireddu points out that Mantegazza's interest in the project of "making Italians" emerges in the pages of this later novel as it explores "the role of the State in the management of new territory, to the importance of improvements in public health, preventive medicine, and factory conditions (in the year 3000 child labor has been abolished [...]), the standardization

As Mantegazza was completing *3000* in 1896, the same year as Italy's disastrous colonial defeat at Adwa, he published a memoir containing parliamentary addresses, personal recollections, and political commentary titled *Ricordi politici di un fantaccino del parlamento italiano* (Political Memoirs of a Parliamentary Foot Soldier). A proponent of Italian colonialism in the Horn of Africa, Mantegazza had been Italy's representative at the Berlin conference in 1884. He was discouraged by the Italian defeat at Adwa in 1896 and by what he saw as an inept colonial government.[11] In his memoir, he transcribed a speech that was intended for his fellow deputies in the Camera some years before in which he called for public health legislation that would remedy the "expiration of the Italian race":

> Sopra 17 dei principali Stati europei, tutti, meno la Russia, godono di un *grado di vitalità che non è concesso al nostro paese*. Il sorriso del nostro cielo, la mitezza del clima, l'amenità dei colli per vendemmia festanti, e le mille altre vanterie di cui risuonarono a lungo gli idillii dei nostri poeti, hanno un riscontro piuttosto doloroso in codesto manifesto *scadimento della razza italiana* originato dalla lunga dissuetudine da ogni virile esercizio, e *alle cui forze vitali* non si cerca mai di apprestare nessuno di quei provvedimenti sanitari per cui vanno lodate le altre nazioni civili. Dopo la Russia noi siamo dunque gli uomini che più si ammalano e che più muoiono in Europa. (*Ricordi politici* 33–34, emphasis mine)

> [Of the more than seventeen principal European states, all except for Russia enjoy a *level of vitality that is not permitted our country*. The smile of our sky, the mildness of the climate, the amenity of the hills for festive harvests, and the thousands of other boasts that resound in the idylls of our poets have been painfully put in check by this obvious *decline of the Italian race*, which began with a gradual desuetude of every virile exercise; *[a race] whose vital forces* we never seek to shore up through public health measures for which other civilized nations are praised. After Russia, we are therefore the men who get sick and die more often than in all of Europe.] (emphasis mine)

and radical reform of the educational system, and the need for higher moral and aesthetic norms in Italian life" (*The Year 3000* 28–29).

11 On Mantegazza's relationship to colonialism, see: Labanca, "Un nero non può essere un bianco" and Choate, *Emigrant Nation*.

The stakes of Mantegazza's public health discourse, which he disseminated in the form of popular hygiene handbooks, manuals, and novels, are thus laid bare as he laments the lack of "vitality" of the Italian race. Making Italians *vital subjects* is arguably at the center of the Mantegazzian oeuvre. For Mantegazza, "Italians had not just to be 'made' politically into a *polis*, but also had to be 're-made' biologically, by science and medicine, into a healthy and vigorous nation," writes Maria Sophia Quine (141). Racial scientists such as Mantegazza, and his colleagues Giustiniano Nicolucci (1819–1904), and Giuseppe Sergi (1841–1936) viewed Italians, as Quine puts it, "as a work in progress, a living, organic mass of bones, bodies, and brains to be skillfully crafted into a *popolo-nazione*" (152). The passage above dwells once again in the semantic orbit of the rhetoric of loss, as Mantegazza's warning about the "decline" of the Italian race presupposes an originary racial unity—itself the product of ideological fantasy. Rhetorical "decline," as a loss of racial prestige or physiological quality, thus functions according to the logic of fetishistic disavowal, insofar as it allows the text to claim a presence—"the Italian race"—in spite of its absence.

L'Anno 3000 weaves the reproductive nucleus of *Un giorno a Madera* together with a more sustained (and arguably more urgent) meditation on the optics and microscopics of race. The earlier *Madera* is a novel about failed reproduction that, given its epistolary structure, foregrounds writing as prophylaxis. The preserved letters that make up the novel are a testament to the fact that no final union between Emma and William takes place and that, therefore, no biologically unfit being is born. As long as Emma and William continue to write, there is no risk of contact, nor of the diseased reproduction that would purportedly result from it. In *3000*, Mantegazza shifts his focus to technologies of racial visualization by projecting a future in which skin is rendered "as transparent as glass." Though the novel grapples with the possibilities of both visible and invisible hereditary markers, *3000* ultimately reaffirms technologies of racialization that rely upon the apprehension of race by the human eye.

Capturing the Moribund Race

Mantegazza's shift from a novel that foregrounds writing, race, and reproduction to one that explores instead the visual parameters of race and reproduction might be explained by his pioneering work in the nascent field of visual social sciences during the second half of the nineteenth century, which was facilitated by advances in photography and Mantegazza's faith in the photograph as a "precious aid" to social scientific discovery. These were the years of Eadweard Muybridge's and Étienne-Jules Marey's experiments with human and animal locomotion that led to the diffusion of stop-action and chronophotography. In Mantegazza's day, Florence was an important site in the development of the photographic medium as both an artistic and scientific technology, as well as a touristic commodity. In 1852, before Italy came into being as such, brothers Leopoldo, Giuseppe, and Romualdo Alinari opened one of Italy's first commercial photographic studios, specializing in portraiture, landscape, and monument photographs which were among the first to depict Italian cities (particularly their native Florence) as commodified objects for touristic consumption.[12] Some forty years later, the Italian Photographic Society was founded at the University of Florence (1889). Between publishing the two novels *Un giorno a Madera* (in 1868) and *L'Anno 3000* (in 1897), Mantegazza was named the Society's first president. During these years, Mantegazza had begun using photography to supplement his studies of the "physiognomy of pain," as well as his anthropometric studies in Lapland and India. Physiognomy—coined in the late 1770s by Johann Caspar Lavater to describe the science born from "the original language of nature, written on the face of Man" (Sekula 10–11)—assigned character traits to anatomic features of the face, thus making of the body a legible text that even a non-specialist might read or interpret. Allan Sekula has cogently identified an "archival paradigm" at the origins of photography's use for social regulation and discipline. When operating in an archival mode, photography seeks to "encompass an entire social terrain while positioning individuals within that terrain" according to types—"heroes, leaders, moral exemplars, celebrities, and those of the poor, the diseased, the insane, the criminal, the nonwhite, the female, and all other embodiments of the unworthy" (10). Through a close reading of two late nineteenth-century thinkers who employed photography to social-statistical ends—Alphonse Bertillon (1853–1914), credited with inventing the first modern system of criminal identification, and Francis Galton (1822–1911), the notorious founder of eugenics—Sekula uncovers physiognomy, and the related sciences of phrenology and criminology, at the "ignoble" origins of realist photography—an instrumentalist and utilitarian practice aimed at arresting the entire social field.

12 See Pelizzari, *Percorsi della fotografia in Italia* 50–52. Pelizzari's book was originally published in English as *Photography and Italy*.

In 1876, as Bertillon and Galton were busy developing photographic methods of classification through the photographic and textual criminal identification card and the photographic composite, respectively, Mantegazza collaborated with photographer Giacomo Brogi (1822–1891) to publish the *Atlante della espressione del dolore* (Atlas of Expressions of Pain). The inspiration for the *Atlas,* along with his subsequent *Fisionomia e mimica* (Physiognomy and Expression, 1881), was Charles Darwin's *Expression of Emotion in Man and Animals* (1872)—one of the first scientific volumes to contain photographic illustrations (M. Pasini 127). The majority of the hundred or so photographs in Mantegazza's *Atlante* were of major works of art representing human suffering, and only a handful were of human subjects on whom various procedures intended to cause physical or sensorial pain were carried out—having their fingers crushed for brief or prolonged periods of time, chewing on bitter wood, and listening to the sound of fingernails scratching glass (which one can imagine was akin to hearing fingernails on a chalkboard), to name just a few. That Mantegazza photographed works of art alongside human subjects also reveals something about the content of physiognomy in the mid-to-late nineteenth century: like Mantegazza's oeuvre as a whole, physiognomy "provided a discursive terrain upon which art and the emerging bio-social sciences met during the middle of the nineteenth century," as Sekula puts it (23). Mantegazza's *Atlante,* a fascinating text, undoubtedly contributes to Italy's anthropological tradition of visualizing, indeed fixing, racial difference through photography, which is most often associated with the criminological studies of Bertillon and Galton's contemporary and Mantegazza's interlocutor, Cesare Lombroso (1835–1909). In the *Atlante,* Mantegazza's subjects are either presented as either presumably neutral or "Negro" subjects; captions reading "Expression of olfactory pain," appear just before "Negro expression of olfactory pain."

Capturing race was also the aim of Mantegazza's two other forays into photography during his trips to Lapland in 1879 and a few years later to India. During his research stay in Lapland, Mantegazza, along with photographer and travel mate Stefano Sommier, used photography to produce a racial archive of the indigenous Lapp people (Fig. 2.2).

Mantegazza's goal was to preserve for "future memory" the Lapp's relative isolation from other "races" such as Norwegians, Swedes, and Russians, since their isolation was bound to end, producing in them what he described as a "racial" change (Mantegazza, *Un viaggio in Lapponia*). Similarly, with his voyage to India, he sought to create a visual archive of the "moribund" pastoral Toda people of southern India (Mantegazza, *Studi sulla etnologia dell'India*).

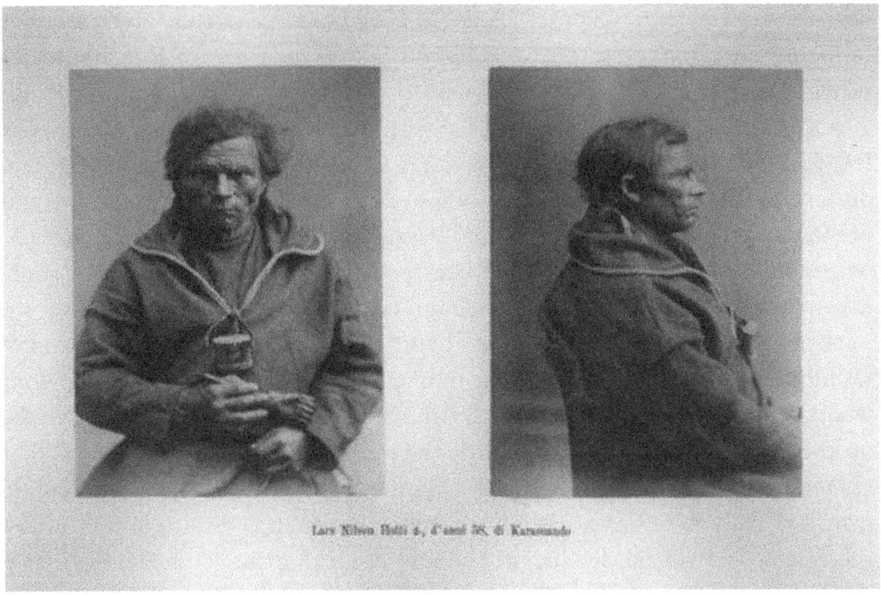

Fig. 2.2 From Mantegazza, *Un viaggio in Lapponia coll'amico Stefano Sommier* (1881)

Prompted by his enthusiastic reception of phrenologist William Elliot Marshall's 1873 study *Travels amongst the Todas*, Mantegazza set out to verify the craniological, psychological, and physiognomic uniqueness of this "race," which he classified as, to paraphrase, closer to Jews and the Semitic races than to the Aryan races. In order to confirm his findings, he took a series of portraits (head shots and profiles), as well as some ethnographic photos of the daily life and customs of the Toda people, and, for comparison, photographs of other groups he encountered during his travels (Hindus, Tibetans, and Lepcha). A true *fin-de-siècle* positivist, Mantegazza believed in the objectivity of the photographic record, in its ability to reproduce reality with scientific accuracy. For Mantegazza, the photograph did not substitute or transcend scientific research, but served instead as a visual supplement to it. Ever captivated by discovering new means to transmit scientific knowledge to the popular masses, Mantegazza also saw photography as a powerful instrument of "democratic" diffusion (*divulgazione*).[13] These accounts of his use of the visual medium to document and produce racial difference provide an interesting explanation for

13 For my discussion of Mantegazza's anthropological photography, I draw from Chiozzi, "Fotografia e antropologia nell'opera di Paolo Mantegazza." See also Chiarelli.

his shift from "writing as prophylaxis" in *Un giorno a Madera* to "technologies of seeing" in *L'Anno 3000*.[14] I locate Mantegazza's turn to visual technologies of racialization in this later novel within the context of his enthusiasm for anthropological photography aimed at verifying scientific "truths" about race. In addition, it bears noting that his scholarly commitment to protecting and preserving life (in the form of public hygiene manuals, novels, essays, and photography)—from the "moribund" Lapps and Todas to Italians themselves, who, as Mantegazza pointed out to his parliamentary colleagues with some urgency, risked racial "decline"—is haunted by imminent death. As Roberto Esposito has suggested, "[T]he living being begins to enter into the horizon of visibility for modern knowledge the moment its constitutive relationship with what continually threatens to extinguish it emerges. Sickness and death make up the cone of shadow within which the life sciences carve out their niche" (*Immunitas* 14–15). To contemplate Mantegazza's purported fear of decline from a slightly different angle, his photographic practice is also bound to the rhetoric of loss (racial decline, as the loss of racial prestige or identity); as a means of capturing the "moribund races" of Lapland, the Toda, or Europeans themselves, the photograph is the ultimate medium for establishing a presence in spite of an absence, as both the subject of the photograph and the moment in time captured are forever lost. Yet the rhetoric of loss, as opposed to absence, nevertheless paradoxically affirms an originary presence; for Roland Barthes, the photograph (and in particular the portrait) contains the affirmation: "[this] has indeed been" (115). Such is the fetishistic nature of the Mantegazzian photograph: by affirming the presence of racial identity before it is "lost" forever, the fundamental emptiness of that identity is kept at bay.

Aesthetic Truths and Fictive Science

Like Leopoldo Franchetti, Paolo Mantegazza has been the object of surprisingly little scholarly attention. This "unjustly neglected" founder of Italian anthropology has received far less scholarly attention than his

14 A 2010–2011 exhibition at the National Museum of Anthropology and Ethnology (which Mantegazza founded), titled "Obiettivo uomo. L'antropologia fotografica di Paolo Mantegazza" (curated by Paolo Chiozzi, Maria Gloria Roselli, and Monica Zavattaro) showcased over a hundred of Mantegazza's ethnographic photos from his travels in Lapland and India, and his photographic studies of anthropometry and physiognomy (Zarrilli).

famed interlocutor Cesare Lombroso, to name just one example.[15] As with Franchetti, when scholars have examined Mantegazza's oeuvre, they have been careful to underscore his playfully innocuous patriotism, often contrasting him with his infamous contemporaries such as Lombroso and Giuseppe Sergi. Despite his subscription to biological determinism and his seminal role in shaping nineteenth-century Italian physical and social sciences, as well as visual anthropology, Mantegazza is distanced from these more "fanatic" interlocutors.[16] Such a relegation betrays a critical ambivalence about where to locate this intensely prolific and influential nineteenth-century thinker within the imbricated genealogies of liberal Italian nationalism and racial discourse.

Mantegazza helped give shape to an Italian school of nineteenth-century positivist social science that forged a connection between human psychology and physiology (Papini 11).[17] The scientific "discovery" of the Italian *volgo* or folk was given a decisive push after Unification, when anthropological and ethnographic pioneers such as Giustiniano Nicolucci, as well as Mantegazza and Lombroso, began to study the ethnic bases of "Italic" peoples. Anthropology and related disciplines such as ethnography, demography, and

15 The phrase is Pireddu's in "The Anthropological Roots of Italian Cultural Studies." Nicoletta Pireddu has edited and written extensive critical introductions to the recent English translations of Mantegazza's *The Physiology of Love* and *The Year 3000*. See: Pireddu, "Introduction. Paolo Mantegazza: A Scientist and his Ecstacies"; "Introduction. Paolo Mantegazza, Fabulator of the Future."

16 Historian of Italian anthropology Giovanni Landucci points out that, "He did not spare fanatic Darwinians and he ended up cutting all ties with old mates such as Sergi and Lombroso" (126). For a similar claim, see Pireddu, "Introduction. Paolo Mantegazza, Fabulator of the Future" 32. Horn suggests that this personal and professional split between Lombroso and Mantegazza arose in 1867 after a "very public debate" about the physiology of pain (*The Criminal Body* 92). Mantegazza later predicted the demise of criminal anthropology, calling Lombroso the "founder of a new religion" and a "false prophet." (Horn, *The Criminal Body* 171, n. 48). See also Mantegazza, *Fisiologia del dolore*. For more on Mantegazza's role in shaping nineteenth-century Italian anthropology, see Puccini, *L'Uomo e gli uomini*.

17 Pireddu argues that Mantegazza's relationship to positivism was ambivalent at best ("Introduction. Paolo Mantegazza, Fabulator of the Future" 49, n. 26). Historian of anthropology George Stocking suggests that the relationship between physiology and psychology—which was in wide circulation by 1860—can be traced at least to Herbert Spencer's *Principles of Psychology* of 1855 (*Victorian Anthropology* 142). Mantegazza's status as an unsuccessful precursor to Freud seems secondary to Alberto Capatti, who faithfully records the spoils of Mantegazza's commercial success in order to trump his famous successor (10).

folklore studies thus emerged and gained momentum in a post-Unification context that was eager to address the fragmentary nature of Italians with regard to class, language, relationships to capitalist modernity, and ethnic belonging.[18] For Pireddu, Mantegazza sought to address the fragmentary nature of Italians first and foremost through a "democratic" educative campaign aimed at "the common reader in its broadest sense" ("Anthropological Roots" 82). Mantegazza also carved out a role for science in the management of the state, and he took a particular interest in female reproductive health. In 1871, thanks to partial funding from the Ministry of Agriculture, Mantegazza (aided by Lombroso), distributed ten thousand copies of a questionnaire to every municipal government in the country to gather information on the *stirpi italiche*, which took special aim at the reproductive health of young women: in addition to height, age, blood pressure, diet and nutrition, the survey queried respondents on the age of menarche. Mantegazza worked for eight years to compel municipalities to complete the survey; he was eventually forced to abandon the project due to the lack of response (Quine 142–143).

Mantegazza was also an immensely successful novelist whose image was once reproduced on matchbooks (Labanca, "'Un nero non può essere un bianco'" 82).[19] For roughly fifty years between 1860 and his death in 1910, Mantegazza waged what Monica Boni has called a "publishing crusade," producing a staggering number of articles, novels, and monographs on bourgeois moral and physiological hygiene (11). As the self-proclaimed "in-house doctor," Mantegazza published popular serials such as the "Almanac of Popular Hygiene" and "The Small Library of the Italian People," which aimed to bring medical knowledge about household management and personal hygiene inside the domestic walls (Quine 140). Unlike his contemporaries Nicolucci, Sergi, and Lombroso, who studied cranial morphology with an eye toward establishing the origins and hereditary transmission of 'Italic' types—whether "Aryan," "Mediterranean," or "degenerate"—Mantegazza looked to anthropometric data as the basis of political

18 See Puccini, "Giustianiano Nicolucci." For connections between the southern question, colonialism, and the disciplines of demography and folklore studies, see Lombardi Satriani, "Realtà meridionale e conoscenza demologica."

19 Pireddu highlights Mantegazza's interdisciplinarity, as well as his status as a forerunner of cultural studies, by quoting his introduction to American audiences as "Physician-surgeon, Laboratory-experimenter, Author-editor, Traveller-anthropologist, Professor, Sanitarian, Senator" ("Introduction. Paolo Mantegazza, Fabulator of the Future" 2). See also Pireddu, "Introduction. Paolo Mantegazza: A Scientist and His Ecstasies" 7.

intervention.[20] Whereas his primary concern was with crafting a biopolitics that served the nation-state, Mantegazza's influence was not limited to the Italian peninsula. Although he was simultaneously revered and rebuked by writer, critic, and cleric Giovanni Papini as a (failed) precursor to Freud and Havelock-Ellis, Mantegazza was an interlocutor of both, along with Richard von Krafft-Ebing, Charles Darwin, Max Bartels, Francis Galton, and Paul Bartels (Pireddu, "Introduction Paolo Mantegazza, Fabulator of the Future" 2).

Despite Mantegazza's pioneering contributions to a number of fields—he experimented with artificial insemination and skin grafts; he is credited with having imported the first coca leaves to Europe and conducted the first scientific studies of the effects of erythroxylon on the human body; and he was an early promoter of the therapeutic bath—several of his contemporaries offered condemnations of the dubiousness of his sexual science.[21] Liberal Italy's preeminent philosopher Benedetto Croce (1866–1952) described allegations among nineteenth-century readers that Mantegazza "abus[ed] science to satisfy non-scientific curiosities" and exploited science as a veil for his own lasciviousness in his study *Gli amori degli uomini* (The Sexual Relations of Mankind, 1886) (Croce, "Scienziati-letterati" 52). Similarly, writer and critic Giovanni Papini (1881–1956) dubbed Mantegazza "The Erotic Senator," characterizing him as a second-rate plagiarist whose publications simply showcased a string of erotic anecdotes (11). Mantegazza's texts were thus the alleged outcome of his prurience. They also had the capacity to inspire dangerous "phantasies" in their female readers. Suspicion was cast on Sigmund Freud's Dora, famed protagonist of his foundational case study on hysteria, by Herr K. when he suggested that she "took no interest in anything but sexual matters," citing her reading of Mantegazza's *Fisiologia dell'amore* (Physiology of Love, 1873) and "books of that sort" as proof that she had invented his indecent proposal to her on a lakeside walk (Freud, *Dora* 41).[22]

While *fin-de-siècle* readers such as Croce, Papini, and Freud stressed

20 For this reason, Mantegazza has been figured as a precursor to Francis Galton (1822–1911) and Alphonse Bertillon (1853–1914) (Quine 142).

21 For a more extensive discussion of Mantegazza's contribution to these and other fields, see Pireddu, "Introduction. Paolo Mantegazza: A Scientist and His Ecstasies"; and "Introduction. Paolo Mantegazza, Fabulator of the Future" 2.

22 For more on Mantegazza's relationship with Freud, see Pireddu, "Introduction. Paolo Mantegazza: A Scientist and His Ecstasies" 8–9.

Mantegazza's texts' erotic potency, several post-World War II readers of Mantegazza share instead a preoccupation with his subscription to Darwinist racism. This recent scholarship, confined less explicitly by bourgeois norms of sexual propriety, is forced to reckon instead with the author's firm entrenchment within the racist epistemologies of *fin-de-siècle* positivism. For instance, in her article problematizing Mantegazza's position vis-à-vis positivism and positioning him instead at the origin of Italian cultural and postcolonial studies, Nicoletta Pireddu asks: "How to cope [...] with the embarrassing remarks on women, or on allegedly lower races that emerge from many of Mantegazza's writings, which, despite his allegedly open and tolerant attitude, seem to plunge him back into the most obtuse survival-of-the-fittest logic?"("Anthropological Roots" 83). She goes on claim that it is precisely this aspect of Mantegazza's work that has led to his post-World War II condemnation and neglect by scholars. Pireddu calls for a "diluting" of what she calls Mantegazza's ideological "shortcomings" with regard to his work as a whole, and in particular in light of what she reads as his challenge to post-Unification positivism and bourgeois rationality (83–84). Similarly, Paolo Chiozzi characterizes twentieth-century critics as "hasty" in their accusations of racism based on Mantegazza's "ambiguous" use of taxonomic models ("Esistono gli 'Ariani'?" 46).

If his formulations of sexual and racial discourse are two components of Mantegazza's work that have received heightened attention from scholars, Mantegazza himself engages various articulations of race and reproduction in order to characterize his narrative project as a whole. He figures his own textual production as both biological and prospective, thus forging a connection between race and reproduction that merits interrogation. At stake in each of the interpretations of his nineteenth-century critics is the degree to which writing can inspire, or is inspired by, unbounded sexuality. This node, contemplated from a different angle, is central to Mantegazza's work, as well. In producing a model of sexual difference in *Fisiologia della donna* (Physiology of Woman, 1893), Mantegazza renders biological reproduction analogous to intellectual or artistic production: "The people of Philadelphia believe that with every childbirth, a woman must lose a tooth. At times, this is true for us as well. We really do lose hair, and after giving birth to a book or a statue, we men also lose health and strength" (64). Male (re)production is thus figured as a form of artistic creation.

More explicitly, in Mantegazza's preface to *Dizionario d'igiene per le famiglie* (Family Hygiene Dictionary, 1881), a collaborative work with the novelist Neera, their textual project is figured as a reproductive one:

> Se il nuovo figliuolo sia nato robusto e destinato a lunga vita, toccherà a voi il dirlo. Io non posso dirvi altro se non questo, che fu concepito con caldo amore e salutato dai genitori con molte e dolci speranze. Possa vivere lungamente e veder vivere i nipoti e i pronipoti di una gagliarda e feconda generazione. (ii)
>
> [It is up to you to say if our new son is born robust and destined for a long life. I can only tell you this: he was conceived with affectionate love and welcomed by his parents with many kind hopes. May he live long and see the jovial and fecund generation of grandchildren and great-grandchildren.]

Following Diotima's address in Plato's *Symposium*, physiological parturition is a metaphor for intellectual production.[23] Significantly, though the text is co-authored, only Mantegazza offers a preface. Neera is thus inscribed in the preface only insofar as she is figured as a (textual) mother. Furthermore, her contributions to the dictionary are marked with an "N.", while Mantegazza's remain unmarked. *Auctoritas*, etymologically linked to author, authority, and "planter-cultivator," is thus doubly aligned with paternity, both in the presence of solely Mantegazza's prefatory comments and in the absence of the need for an authorial signature on his dictionary entries. The newborn text is positioned at the origin of a prospective male genealogy; Mantegazza thereby promises readers ever more textual offspring.

If, as the selections from Mantegazza's *Fisiologia della donna* and *Dizionario d'igiene per le famiglie* cited above illustrate and as the nineteenth-century critics' concerns confirm, Mantegazza's textual project is figured as (sexually) reproductive, race comes to the fore, given the historical intersection between the production of sexuality and biopolitical race thinking that Foucault famously identified. Foucault contends that the bourgeois deployment of sexuality was the result of a shift from an aristocratic "symbolics of blood" to a bourgeois "analytics of sexuality." While within narratives of aristocratic

23 Mantegazza arguably biologizes Plato's metaphor. While intellectual and/or moral formation is at stake for both, Mantegazza's project is one that draws narrative and physiology, specifically with regard to biological progeny, into closer proximity. For a discussion of Diotima's address that examines the multiple forms of spiritual and physiological pregnancy operating therein, see Pender. Adriana Cavarero in "Diotima" offers a reading of Diotima (as performed by Socrates) as a privileged and necessarily female exponent of Plato's philosophy.

nobility, blood served a retrospective function (genealogy), the emergence of a bourgeois discourse of sexuality was centered around the prospective function of blood (racialized progeny). "The bourgeoisie's 'blood,'" claimed Foucault, "was its sex" (*The History of Sexuality* 128). Reproductivity is thus the discursive site upon which individual bodies are inscribed in the service of a racialized collectivity.

Instead of attempting to embellish Mantegazza's racial politics by deeming them more or less virulent than his contemporaries or successors, or relatively anomalous or insignificant when compared to his greater body of work, I propose instead a reading of how Mantegazza figures the Italian national body as made up of *vital subjects* and how other bodies (pathologized, sexualized, racialized, etc.) perform boundary work, ensuring the (fantasized) cohesiveness of that body. Following Stuart Hall and Paul Gilroy, the question passes from "Is it racist?" to "How does race work?"[24]

One way to address the question of how race works in the Mantegazzian text is by exploring how he simultaneously exploited the registers of art and science. For Chiozzi, Mantegazza's unfavorable response to Austrian anthropologist Karl Penka's 1883 article on Aryan origins is proof that Mantegazza was not racist. Mantegazza writes:

> Gli Ariani sono ancora per noi un mito storico in cui il vero si associa a molta nebbia e fors'anche a molti errori. Io leggo con molta attenzione tutto quanto si scrive sulle origini dei popoli, ma fino ad ora non trovo che romanzi storici nei quali mi è assai difficile il discernere quanta parte spetti alla storia e quanta alla fantasia dell'autore, e senza dire il troppo superbo ignorabimus, chiudo il libro, mormorando modestamente: ignoramus. (*Fisionomia e mimica* 364)

> [Aryans are still a *historical myth* for us wherein what is true is foggy and perhaps contains many errors. I have read quite attentively everything that has been written on the origins of peoples, but until now I have been unable to find anything but *historical novels,* in which it is quite difficult for me to discern how much belongs to history and how much belongs to the imagination [*fantasia*] of the author, and, without saying

24 Here I draw from Stuart Hall's call for a focus on the "concrete historical 'work'" performed by racism, and Paul Gilroy's suggestion that anti-racism inhibits our ability to view race as a "precarious discursive construction" (Hall; Gilroy, "The End of Anti-Racism").

> the overly arrogant "ignorabimus," I close the book, modestly mumbling "ignoramus."] (emphasis mine)[25]

Here, Mantegazza anticipates contemporary critiques of race that figure its production as a battle between "truth" and "fiction." For Mantegazza, authorial fantasy is operative in ensuring that origin stories—in this case that of the Aryans—remain within the dual realms of myth and ignorance.

Yet in Mantegazza's *Fisionomia e mimica*, Aryans, as distinguished from Semites, take center stage as representatives of the European races:

> Hanno la pelle bianca quasi tutti gli Ariani e i Semiti e a quanto dicesi molti polinesiani, che non sono né malesi, né papuani e che probabilmente hanno comunque con noi la stessa origine. Hanno la pelle nera i Negri, i Papuani, gli Australiani, alcuni tribù dell'India e i Negriti. Tutti gli altri popoli della terra hanno la pelle del color delle fave secche [...] (38).

> [Almost all Aryans and Semites have white skin, as do, from what I hear, many Polynesians who are neither Malaysians nor Papuans, and who probably have the same origin as we do. Negroes, Papuans, Australians, and some tribes of India, as well as Negrites have black skin. All other peoples on earth have skin the color of dried fava beans.]

These categories are deployed in order to hierarchize:

> In alto e in basso dell'albero umano rami e ramoscelli si avvicinano, per modo che altissimi e bassissimi si toccano. Il negro che si eleva a *cafro* si avvicina all'europeo, che col gozzo o il cretinismo o la fame si abbassa, si avvicina all'australiano e al negro (93).

> [High and low on the tree of man, branches large and small approach one another, so that the very highest and the very lowest touch. The Negro who is elevated to the level of Kafir approaches the European, who, thanks to a goiter, idiocy, or hunger, is lowered, approaching the Australian or the Negro.]

Mantegazza's attack is therefore waged not at the Aryan as a category, but

25 For a more nuanced (yet largely biographical) discussion of Mantegazza's shifting relationship to anthropological race, see Labanca, "'Un nero non può essere un bianco.'"

at the speculative nature of origin stories in general: "In classifying races [*razze*] we must exclude their origins as much as possible, because searching for origins is the most fecund source of ethnologic errors" (93). Mantegazza thus upholds the (fictive) truth of the Aryan while condemning speculations about his provenance to the realm of error.

Furthermore, if Mantegazza disparages the fiction of origins stories in the name of truth, in another instance, he complicates this reading by assessing the function of fiction in scientific discourse:

> [I]l vocabolo razza [...] non è che il prodotto della nostra ginnastica mentale, delle operazioni del nostro intelletto, all'infuori di qualsiasi realtà. La scienza aveva bisogno delle razze come di quadri ipotetici e questi prodotti dell'arte, per servirci di un'espressione di Lamarck, sono diventate per il volgo realtà concrete. Le razze esistono come una finzione del nostro cervello; esse esistono in noi, ma non fuori di noi. (Cited in Chiozzi, "Esistono gli 'Ariani'?" 47)
>
> [[T]he word *race* [...] is nothing but a product of our mental gymnastics, the operations of our intellect, outside of any reality. Science needed races [to serve as] hypothetical frames, and these products of art, to use a Lamarckian expression, became concrete realities for common folk. Races exist only as an invention of our brains; they exist within us, but not outside of us.]

Mantegazza draws from Lamarck's 1809 treatise *Philosophie zoologique* in order to echo the latter's assertion that classification is a mere interpretive apparatus ("hypothetical frame" or "product of art"), adopted by the naturalist in order to make sense of his object (nature).[26] Mantegazza extends Lamarck's analysis to describe not only the task of the scientist (deploying "art" to organize the natural world), but also the way in which scientific methodology

26 "Le but, au contraire, d'une classification des animaux est de fournir, à l'aide de lignes de séparation tracées de distance en distance dans la série générale de ces êtres, des points de repos a notre imagination, afin que nous puissions plus aisément reconnaître chaque race déjà observée, saisir ses rapports avec les autres animaux connus et placer dans chaque cadre les nouvelles espèces que nous parviendrons à découvrir. Ce moyen supplée à notre faiblesse, facilite nos études et nos connaissances, et son usage est pour nous d'une nécessité indispensable; mais j'ai déjà montré qu'il est *un produit de l'art* et que, malgré les apparences contraires, il ne tient réellement rien de la nature" (emphasis mine; Lamarck 117).

Fig. 2.3 *Fisiologia del piacere* (1854)

permeates (however erroneously) public consciousness ("[races] became concrete realities for common folk"). In so doing, Mantegazza exposes race as a discursive production, without, however, abandoning its organizing function in scientific narrative.[27]

In staging a critique of race, Mantegazza simultaneously if implicitly addresses his own textual production. Following Lamarck, he claims that science relies upon "art" or "fiction" for its interpretive frame. He thereby authorizes his own role as a writer of (science) fiction. Furthermore, by evoking the misled "common folk" (*volgo*) he hails the very audience he seeks to infuse with the morally hygienic knowledge through his popular prose. Mantegazza's texts are a showcase of meditations on the multiple articulations of science and aesthetics (Pireddu, *Antropologi alla corte della bellezza*). His "scientific" texts brim with naturalist metaphors, poetic hyperbole, and extended digressions on beauty.[28] Taken together, his many moral treatises entitled "Physiology of [Love, Pleasure, Pain, etc.]" might be said to constitute a literary genre unto themselves (Fig. 2.3).

27 Far more important than racial classification for Mantegazza's scientific project was a biological explanation of inheritance. Darwin would praise Mantegazza's work on pangenesis as an explanation for the transmission of both inherited and acquired traits as extremely influential in his groundbreaking study of evolution (Taylor 10).

28 This earns Mantegazza the distinction of being more of a *noveliere* than a scientist for Sandra Puccini ("I viaggi di Paolo Mantegazza" 51).

Conversely, his novels have typically been read as artless vehicles for his scientific (or less-than-scientific, following his scandalized detractors) agenda.[29] Clearly, one of the central tensions within Mantegazza's oeuvre is an oscillation between the registers of science and literature in the service of his global taxonomic mission to render psychological and physiological man—and woman (and herein lies the threat of his work, at least to nineteenth-century readers)—legible and intelligible.[30] If race is more artful fiction (narrative) than science, it is necessary to both his scientific and literary projects. Indeed, Lucia Re has referred to the porous boundaries between literature and the Italian racial theory of Niceforo and Sergi in this period in terms of "rhetorical short circuits" ("Italians and the Invention of Race" 20).

A fervent Darwinist, Mantegazza availed himself not only of vocabularies of zoological typologization, but also of logics of racialized and nationalist hierarchy.[31] True to *fin-de-siècle* form, in Mantegazza's writings, at times the national population contains (by subsuming) racialized variation, and at others the nation stands in metonymically for race, or vice versa. This blurriness recurs throughout Mantegazza's writings; the shifting logic is typically nestled within a rhetorical oscillation between humanistic univer-

29 Walter Pasini resists classifying Mantegazza's novels as such: "He was a writer of books of a novelistic character. [...] It should be said that Mantegazza did not achieve excellence in this field because his novels have far too obvious didactic intentions and autobiographical notes that were a bit too insistent" (18). This condemnation brings Mantegazza closer to his colleague and pen-pal, Charles Darwin: "For [historian Walter Cannon] the *Origin* is a dull monograph whose amateurish metaphors (a tree, a bank, a struggle, a chain, a beehive) are superfluous decoration on a structure built out of the massive freestone blocks of documented fact. These figures of speech, he says, are no more than 'available verbalisms' and Darwinism itself profoundly anti-literary because it proves conclusively in its style and its freedom from moralistic assumptions that science has outgrown all the ancient resources of myth, ritual and drama" (Morton 5).

30 This oscillation was also institutional: the chair in Anthropology that Mantegazza occupied beginning in 1869 (Italy's first) was originally housed in the University of Florence's Facoltà di Filosofia e Lettere. Mantegazza appealed relentlessly to the administration for its transfer to the Facoltà di Scienze (Landucci, 113; 126).

31 Mantegazza was cited by Darwin in *On The Origin of Species* (1859) and *The Descent of Man, and Selection in Relation to Sex* (1871). Mantegazza also dedicated several lectures and articles to Darwin's theories and their reception. See for example: "Commemorazione di Carlo Darwin. Discorso di Paolo Mantegazza." Florence: Arte della Stampa, 1883 and "Darwin dopo 50 anni," A.A.E., vol. XXXV, 1905; 311–322 (cited in Landucci, 128). See also Pireddu, "Introduction. Paolo Mantegazza: A Scientist and his Ecstasies" 7.

sality ("the human family") and racialized specificity ("high races" and "low races").[32]

In *Fisionomia e mimica,* Mantegazza addresses the concept of race explicitly, writing:

> La razza è un'espressione molto larga, che abbraccia tante e diverse cose, quali un certo modo di sentire, un certo grado di intelligenza, certa intensità di emozioni, e tutte queste cose influiscono e modificano la mimica. È questo uno dei punti più oscuri della mimica e noi dedicheremo ad esso uno speciale capitolo. (294)
>
> [Race is a broad expression, that embraces many diverse things, including a certain way of feeling, a certain intelligence level, a certain intensity of emotions, and all of these things influence and modify facial expressions. This is one of the most obscure aspects of human expression and we will dedicate a special chapter to it.]

In this formulation, as in many that attempt to provide a definition of "race," causality is ambiguous at best.[33] As Mantegazza attempts to provide a definition (it is introduced with the classificatory heading "Race," and follows similar ones such as, "Education" and "Sex"), he opens by anticipatorily foreclosing the task at hand ("race is a broad expression"). He then goes on to list qualities that race might either describe or determine ("a certain way of feeling, a certain intelligence level, a certain intensity of emotions [...]").

32 This relationship between specificity and universality is fundamental to Victorian evolutionary anthropology, which resolved this potentially paradoxical disjunction with temporality. Stocking writes, "mankind was one, not because it was everywhere the same, but because the differences represented stages in the same process" (*Victorian Anthropolgy* 225).

33 Ann Stoler has pointed out that such an opacity of causality is inherent to race thinking. In the epilogue to her aforementioned study *Race and the Education of Desire,* Stoler discusses what Foucault called the "polyvalent mobility" of race by examining the multiple implications within contemporary political discourse of the statement, "Blacks are poor because they are black." Depending upon ideological exigency, the statement may be wielded by self-proclaimed racists (biology determines racial inferiority, which determines class) and anti-racists (institutional racism both produces and codifies race, which determines class) alike. Though Stoler focuses upon the mobility of this declaration when it is adopted by antagonistic political groups, her discussion calls forth the need for the problematization of causality in analyses of racialist discourse.

Rather than answering the question he has posed, Mantegazza defers the reader to a subsequent chapter.

In Mantegazza's meditation on race in *Fisionomia e mimica*, the mobility of the term is evident precisely insofar as causality is muddled. The moral-psychological categories of sentiment and intellectual capacity that for Mantegazza dictate physiognomy both disclose racial truths ("They are racially marked and so they are emotionally/intellectually inferior") and provide the necessary conditions for racialization ("They are emotionally/intellectually inferior and so they are racially marked"). Indeed, Mantegazza's next move is to assert the "obscurity" of the term, promising an entire subsequent chapter aimed at its exegesis. Yet what he delivers merely affirms this obscurity, and recalls his understanding of race (cited in Chiozzi) as a mobile hermeneutic mechanism. Taking issue with climatological analyses of character, Mantegazza privileges instead the influence of racialized blood:

> Gli Scandinavi sono molto parchi nei loro movimenti, poco vivaci, molto silenziosi; hanno in tutto una mimica piena di riserva, io direi, molto concentrica. Ma voi andate a Bergen, una delle maggiori città della Norvegia e vedete invece gente gaia, rumorosa, una mimica eccentrica e tumultuosa. Ma dunque? Anche qui fa freddo e perché la mimica è tanto diversa da quella che si osserva a Trondiem o a Cristiania? È perché a Bergen vi fu nei secoli lontani una grande importazione di schiavi irlandesi; quindi avete molto sangue celtico, che ha portato seco la telegrafia dei gesti e la prorompente vivacità della mimica. E chi mai oserebbe parlare d'una mimica italiana, quando è così diversa a Napoli e a Milano, a Cagliari e a Torino? (296)
>
> [Scandinavians are very frugal in their movements, which are lifeless and silent; they have overall a very reserved, I would almost say concentric, range of expressions. But if you go to Bergen, one of the largest Norwegian cities, you will find instead a joyous, noisy people with an eccentric and tumultuous range of expressions. And so? It is cold there, as well, so why would human expression be so different from that of Trondheim or Christiania? It is because in Bergen, centuries ago, there was a major importation of Irish slaves, so there you have a lot more Celtic blood, which brought along with it the telegraphy of gestures and the uncontainable liveliness of expression. And who would ever try to speak about a unified Italian expression, when it is so different from Naples to Milan, from Calgliari to Turin?]

Blood trumps climate, as Bergenians are merely Celts in disguise. Racialized blood is introduced in order to explain away internal difference, as this fragmentation undermines the metonymic substitution of nation for race.[34] Mantegazza extends this explanation to the Italian peninsula, thereby suggesting that blood ensures southern difference.[35] He closes with the query, "And who would ever try to speak about a single Italian expression, when it is so different from Naples to Milan, from Calgliari to Turin?" Evidently, it is Mantegazza himself who dares speak of a unified "Italian expression," and he does so not a paragraph later; the suggestion of anything less risks destabilizing his fiction of European racial supremacy (and his positioning of Italians within this group). Perhaps not surprisingly, then, his next move is to address this potential for racialized variation within the naturalized borders of the nation-state by producing a model of Italian/black opposition:

> La mobilità dei lineamenti è assai diversa nelle diverse razze e non si accorda sempre col grado di gerarchia psichica. [...] Ma se i negri fanno colla loro faccia molte smorfie, hanno il volto mobilissimo anche gli Italiani, che pure stanno a un livello molto più alto. (297)
>
> [The range of facial expressions is quite diverse within the different races, and it is not always in line with [where a group sits in] the psychic hierarchy. [...] If Negroes make many faces, Italians also have very mutable faces, even though they are at a much higher level [in the psychic hierarchy].]

The contingent threats of internal Italian fragmentation and Italian-black likeness are thus abated as Italians are inscribed at the top of the chromatic hierarchy. The logic of this progression reads: racially fragmented Italians may be, but in opposition to the "black race" (whose geographical coordinates need not be specified), they constitute a unified representative of the "Italian race." Indeed, as Mantegazza was drafting *Fisionomia e mimica*, early Italian

34 It is of further note that in this passage, the human eye deceives and Mantegazza must intervene in order to penetrate the immediately visible with an affirmation of the microscopics of blood.

35 Mantegazza's choice of the oppositional pairs of Naples and Milan and Cagliari and Turin are unequivocal shorthand for southern question discourse that posits geographic extremity as representative and/or productive of cultural, racial and/or linguistic difference.

colonial enthusiasts found the African/European or black/white opposition was the most effective way to erase racialized regional difference (Wong 89). My goal here is not to trace the variations within Mantegazza's definitions of race over the course of his hyper-prolific career, nor is it to highlight the fact that these shifting definitions are often at odds with one another. Such a reading might produce a catalogue that would, for one, reproduce the classificatory logic of Mantegazza himself, and, moreover, would risk engendering little more than the conclusion that his scientific method, like that of so many of his contemporaries, was faulty.[36] Suzanne Stewart-Steinberg identifies a similar tendency in Lombroso criticism: "[his] style—his bad writing, his lack of system, his obsessive accumulation of 'facts'—turned him into an easy target of a debunking that made him the propagator of a pseudo-science whose basic presuppositions could not hold up to closer inspection" (232).[37] The same might be said for Mantegazza. Contenting oneself with condemning Mantegazza's outdated scientific method would lead to a similar dead-end in terms of interpreting his own vast production.

Technologies of Seeing

> Rather than stressing the separation between art and science in the nineteenth century, it is important to see how they were both part of a single interlocking field of knowledge and practice. The same knowledge that allowed the increasing rationalization and control of the human subject in terms of new institutional and economic requirements was also a condition for new experiments in visual representation.
>
> Jonathan Crary, *Techniques of the Observer* (9)

36 As Hannah Arendt argued in 1950, "[T]oday no single science is left into whose categorical system race-thinking has not deeply penetrated" (160). Nancy Leys Stepan reminds us: "[T]he sciences of human difference cannot and should not be dismissed lightly as something belonging thankfully to the past [...] They were the work of the best scientists of the day and were at the centre, not the margins, of science. Evolutionary biology, modern genetics, bacteriology, sciences which still provide the framework for the sciences of biology and medicine today, were all closely tied to racial (and sexual) sciences. Racialism was thus part of the very modernity of science" (31–32).

37 In her study of Lombroso and writer Matilde Serao, Nancy Harrowitz describes one such reading of Lombroso's work as "reductive" and "limited" (23) because it fails to account for his historical and epistemological context.

Though race (*razza*) figures prominently in Mantegazza's "non-fiction" prose, his novel *The Year 3000* is not organized around an explicit deployment of the term. Instead, race appears in the novel as a relic of the past—its representatives are conserved in a futuristic museum exhibit. Yet central to the thematics that structure this fictional text is a project of reproductive engineering. At the opening of *The Year 3000*, Mantegazza promises a futuristic tale aimed at the celebration of the "fertile marriage" (*matrimonio fecondo*) of its protagonists.[38] What the text ultimately produces, however, is a figure of male intellectual production, rather than female parturition. Instead of the birth of a healthy biological being, the "birth" celebrated at the close of the novel is one of scientific genius: the invention of a psychoscope. This invention is the last in a series of technologies of visualization figured throughout the novel aimed at reducing physiological and psychological bodies to a kind of text.[39] Mantegazza's novel thus dramatizes an epistemological shift that occurred in the nineteenth century, and most markedly within his field of physiology, that made the body "the site of both power and truth." As Jonathan Crary suggests in his seminal study of modernity and vision: "[K]nowledge was conditioned by the physical and anatomical functioning of the body, and perhaps most importantly, the eyes" (79). As the relationships between the body, vision, and truth were being refigured, Mantegazza, along with his famed interlocutors Francis Galton, Sigmund Freud, and others, attempted to externalize the mind through technologies of visualization (photography as a "mental picture"). Physiognomy and phrenology might be counted among such technologies of visualization. As Lev Manovich has argued and as Mantegazza's novel makes clear, these attempts grew out of an increasing demand for mass standardization and regulation. Simply put, "The private and the individual are translated into the public and become regulated. What before was a mental process, a uniquely individual state, now becomes part of the public sphere." ("Visual Technologies" 205). For Manovich, modernity addresses subjective interiority through an exterior visual representation of it, paving the way for technologies of mass spectacle such as cinema.

38 In his translation of the novel, David Jacobson translates *matrimonio fecondo* as "mating match" (Mantegazza, *The Year 3000. A Dream* 58).

39 Horn and Stewart-Steinberg both discuss the centrality of corporeal texts to Lombroso's project. Whereas Lombroso was eager to catalogue criminally deviant bodies (indeed, Stewart-Steinberg notes that a void occupies the space of the "normal" subject), Mantegazza was keen to produce healthy bodies as foils to potentially pathological ones. This is particularly the case in what I call his "reproductive novels": *Un giorno a Madera* and *L'Anno 3000*.

Though Mantegazza's psychoscope does not arrive at the proto-cinematic, it is nevertheless pervaded by such an impulse toward the exteriorization of interiority and its regulation (205).

Mantegazza makes clear that his eugenic utopia, *L'Anno 3000* (the least successful of his forays into popular fiction), is in part a response to the central thematic concern of *Un giorno a Madera*—healthy, as opposed to diseased, reproduction: "Couples' visits to be authorized for fertile marriage have fairly well diminished hereditary diseases, but they still exist, through the errors of visiting physicians and through vices that ruin good constitutions as well" (*The Year 3000* 111).[40] In *3000*, protagonists Paolo and Maria travel by electric airship from Rome to Andropolis, capital of the United Planetary States (located at the foot of the Himalayas) to obtain a permit for a reproductive marriage from the Biological Senate.

Premarital certification was, along with birth control, sterilization, and mental hygiene, among the fundamental topics of eugenics debates in late nineteenth-century Italy. Eugenics, coined by Francis Galton in 1883, entered official discourse in Italy in 1912 with the participation of Corrado Gini, Giuseppe Sergi, Alfredo Niceforo, Enrico Morselli, Antonio Marro, Roberto Michels, Achille Loria, and Raffaele Garofalo at the First International Congress of Eugenics in London (Cassata 19; 27).[41] As Mantegazza's writings attest, many of Galton's principles were in wide circulation well before an official Italian eugenics school, the Comitato Italiano per gli Studi di Eugenica, was launched in 1913 (Cassata, *Molti, sani e forti* 27). Indeed, several years before Galton's coining of the term eugenics, Mantegazza published *L'igiene dell'amore* (The Hygiene of Love, 1878), in which he asserted his own primacy over Galton's with regard to studies of heredity (Pireddu, "Introduction. Paolo Mantegazza, Fabulator of the Future" 32; Mantegazza, *L'igene dell'amore*). For Francesco Cassata, Mantegazza's "hygienic-utopian" novels *Un giorno a Madera* and *L'Anno 3000*, alongside other projects of social medicine, and the fleeting appearance of an Italian neo-Malthusian school, all attest to the presence of a strong proto-eugenic discourse in liberal Italy

40 Unlike *A Day in Madeira*, *The Year 3000* saw only one edition, which was re-released by Lumbrina in 1988 with a preface by Alberto Capatti. In 2007, it was released again by the publishing house Lupetti. The novel was published under the series title *I Rimossi*, dedicated to titles that had been "removed" from circulation, or, significantly, "repressed." For an in-depth discussion of the publication, translation, and reception of *The Year 3000*, see Pireddu, "Introduction. Paolo Mantegazza, Fabulator of the Future" 42–44.

41 See also: Galton, *Inquiries into Human Faculty and its Development*. See also Galton, "Hereditary Talent and Character."

("Rigenerare la razza" 115).[42] Mantegazza regarded science as the basis for government policy; Maria Sophia Quine argues that this instrumentalization of science toward explicitly political ends is what distinguished Mantegazza from another of his ethno-anthropological interlocutors, "the 'father' of Italian racism," Giustiniano Nicolucci, and thus brought Mantegazza closer to twentieth-century eugenics (140).

On their way to apply for premarital certification in Andropolis, Paolo and Maria, the protagonists of *L'Anno 3000*, make stopovers in the Necropolis of La Spezia, where Paolo lectures young and inexperienced Maria on the political and military history of Europe since the nineteenth century, and at the Egyptian Pyramids, where the protagonists feast on fish drawn from the sea that now covers the former deserts of the African continent. Paolo and Maria then travel by luxurious, high-speed ocean liner to Sri Lanka (still known as Ceilan in the year 3000), where they visit variously governed city-states: The Land of Equality, Tyrannopolis, Turatia ("The Socialist Republic"), and Logopolis (copy of parliamentary England). From this "Island of [political] Experiments," the travelers continue on to the Island of Dynamo, where they observe the office of planetary energy (whose technicians distribute currents of light, heat, and electricity throughout the globe by way of an elaborate web of conductor cables). From here, they board their *aerotach* once again, bound directly for Andropolis. After visiting the administrative centers of the capital city for a few days, Paolo surprises Maria when he is awarded the Cosmic Prize (conferred by the Academy of Andropolis), for his invention of the psychoscope, which allows its user to read the thoughts of whomever he chooses.[43] Following the ceremony, Paolo and Maria receive authorization to marry from the Health Tribunal. The novel concludes with a telegrammatic account of their wedding.

This felicitous narrative conclusion that I have just outlined does not, however, coincide with the novel's ideological conclusion—in which the

42 See also: Mantovani; Bonetta.

43 Chris Fern notes that technological innovation also plays a central role in the *fin-de-siècle* utopias of Bellamy (*Looking Backward*, 1888) and Wells (*A Modern Utopia*, 1905) and claims that this reflects their historical positioning within an era of "dramatic advances in hygiene, medicine, long-range transport and communications" (69). Nicoletta Pireddu makes a similar argument in her discussion of the literary intertexts and cultural context of *The Year 3000*. For Pireddu, the psychoscope represents the novel's attempt to "extend the moral and social benefits of science even to the less tangible and more complex manifestations of human life—that is, to the psychological and spiritual ones" ("Introduction. Paolo Mantegazza, Fabulator of the Future" 3; 7; 31).

novel acknowledges that the political world it has labored to construct is untenable. Despite the fact that the novel, set in the year 3000, includes an account of the historical surpassing of race, the way that it approaches the relationship between vision and corporeal knowledge reveals that it is not beyond the logics of racialization that prescribe the apprehension of race by the human eye. Central to the couple-protagonists' travels is a visit to the eugenics laboratory known as the Hygeia, where newborns deemed physiologically unfit are incinerated. This marks the final phase of the pair's reproductive education, and enables them to obtain a permit for a reproductive marriage, purportedly the primary aim of their pilgrimage from Rome to Andropolis. Yet this project of eugenic engineering occurs in a future tense which Mantegazza figures as already beyond race:

> Le rapide e facili comunicazioni fra paese e paese e le profonde modificazioni dei climi avvenute per opera dell'uomo tendono ad ogni generazione a fondere indefinitamente le razze, creando un nuovo tipo, indefinitamente cosmopolita, frutto dell'incrociamento intimo e profondo di tante e tante razze, che per lunghi secoli eran rimaste isolate e disgiunte, facendosi paura reciproca e continua e distruggendosi a vicenda col ferro, col fuoco e più ancora col trasporto di terribili malattie infettive, che poi colla cresciuta civiltà sono quasi del tutto scomparse dalla superficie della terra. (146)

> [The rapid and easy communications between one country and another and the profound changes in climate that have come about through human intervention tend in each generation to fuse the races indefinitely, creating a new type, indefinitely cosmopolitan, the fruit of the deep and intimate cross-breeding of ever so many races that had over long centuries remained separate and isolated, creating continual mutual fear and destroying each other with sword, with fire, and even more with the transportation of terrible infectious diseases, which later, with the advances in civilization, have almost entirely disappeared from the face of the earth.] (170)[44]

The "indefinitely cosmopolitan type"—a sort of super-race—can only emerge

44 This fantasy about communication and industry enabling diffusion recurs throughout Mantegazza's oeuvre. In *Fisiologia della donna* he figures the process similarly (186).

in the wake of the disappearance of those which the text has determined are worthy of elimination: "the Australians, the Maori, the Hottentots, the bushmen, many blacks, the Guarani [...]" (*The Year 3000. A Dream* 169). The end of racialization is thereby effected by its mobilization in hierarchized natural selection.[45] The eugenic project of the novel, which might in this post-racial future seem superfluous, thus relies upon a second-order racialization, one in which various medical-scientific technologies are mobilized in order to penetrate the epidermal and physiognomic surface in order to evince physiological and psychological constitution. It is our couple-protagonists' visit to the Museum of Andropolis that signals what I consider to be the ideological conclusion of the novel: Paolo and Maria may only view the mannequins of racialized man on display there by relying upon modes of seeing that the text's technological innovations have rendered obsolete.[46]

In *3000*, travel is pedagogical and is essential to ensuring the eventual biological union of the healthy protagonists. Their educative relationship is likewise gendered: Paolo is a linguistic and (social)-scientific authority; he guides Maria through new socio-political terrains, evoking a crude refiguring of Virgil's pedagogical role in the *Commedia*.[47] As he and Maria prepare for a trip to the offices of the central government, she retreats: "Paolo, dear, I'm an ignorant little woman who finds it hard enough to govern a house and is bewildered at the notion that just a few humans can govern the entire world from the Government Palace of Andropolis" (104). Paolo responds with a "love pat" and authorizes her role as pupil, "No, you are not an ignorant

45 Pireddu argues that here Mantegazza is "in line with the Darwinian theory of evolution, which, by invalidating the notion of the fixity of species, had reconceptualized the life of organisms as subject to continuous and progressive modifications with the passage of time." She also notes that this ethnic homogenization is among the "problematic" moments in Mantegazza's utopian vision. She reads this "new type, indefinitely cosmopolitan" in terms of what she calls Mantegazza's "cosmopolitical democracy" ("Introduction. Paolo Mantegazza, Fabulator of the Future" 30; 39).

46 It is worth noting that Mantegazza's technological innovations are epistemologically rooted in physiology and are consequently described by Mantegazza as prosthetic extensions of the human eye (rather than exterior, objective and/or mechanized agents of sight).

47 Indeed, Paolo has learned the "dead language" of Italian (in 3000, a "cosmic language" is spoken) and he credits Dante as his inspiration (Mantegazza, *The Year 3000. A Dream* 62). For more on the internationalism "cosmic language" of *3000* in relation to Ghislanzoni's European federalist project in *Abrakadabra* (1884), see Pireddu, "Introduction. Paolo Mantegazza, Fabulator of the Future" 38.

little woman, and the central government of Andropolis is not a cabal nor such a dark, intricate mechanism that you can't understand or admire it" (105). This exchange prompts one of Paolo's erudite lectures on the political, military, and social history of the area since the nineteenth century; strung together, lessons like these, aimed at both the reader and Maria constitute the narrative bulk of the novel.[48] Moreover, travel enables Mantegazza to thematize the genre within which his novel might be inscribed: that of utopian fiction.[49] The lands the betrothed must traverse before arriving at their destination are so many utopias, and they are staged as naive and idealistic foils to the text's narrative, thematic, and ideological point of arrival: Andropolis.

Chris Fern notes that one of the fundamental elements of utopian fiction is the protagonist's return "home" and an account of his discoveries ("the protagonist in such narratives is nearly always male") to his fellow (non-utopian) citizens (13). While Mantegazza's couple-protagonists do not return to Rome (the novel concludes in Andropolis), their voyage is staged as the realization of a fictive one of the same name and already completed by another author (who bears a striking resemblance to Mantegazza). At the opening of the novel, the narrator states: "Paolo and Maria brought just a few

48 Curiously, while the protagonists are not engaged in time travel, the narrator—and even at times Paolo, whose learned observations are often indistinguishable from those of the narrator—seems quite conscious that we, as readers, are. That is, instead of orienting the terrain geographically for (primarily) Maria, the narrator regularly orients Maria temporally: rather than "Here [insert sociological fact]," the narrator orients with, "In the year 3000, [insert sociological fact]." Fern observes that the status of the plausibility of utopian fiction shifted between the Renaissance (More, Campanella and Bacon)—which posited geographical elsewheres just as "*actual* travelers' accounts" from voyages of "discovery" were circulating—and the nineteenth century, when time travel became critical to utopian writings, thereby diminishing their plausibility and ensuring the [Fern citing H.G. Wells] "double-encoding of utopias as both fiction and non-fiction, literature and nonliterature." H.G. Wells, *A Modern Utopia*. Lincoln: University of Nebraska Press, 1967; 93 (cited in Fern 71).

49 Though Jameson identifies the homology of socialist and utopian discourse ("Utopia is a transparent synonym for socialism itself," he writes), Mantegazza arguably re-figures this relationship. He stages the socialist communities of Egualianza and Turazia as failed utopias in order to formulate his own—Andropolis—which, he suggests through his condemnation of the oppressive homogeneity operating in these societies, instead purportedly celebrates human diversity. The paradox constituted by such a claim in light of the text's eugenic project is one of the central tensions of the novel. For more on the genres of Italian utopian and science fiction, see Pireddu, "Introduction. Paolo Mantegazza, Fabulator of the Future".

books with them, among them *The Year 3000,* written ten centuries earlier by a physician with a bizarre imagination who tried to guess what human life would be like a millennium on" (*The Year 3000. A Dream* 58). Rather than the account of utopia emerging from the travelers' return home, in *3000,* the protagonists traverse terrain already produced by the fantastic meanderings of a doctor/author. Indeed, one of their goals is to verify the accuracy of this futuristic fantasy. Paolo declares, "I'm really curious how well this prophet guessed the future" (58). Paolo and Maria's "home" is therefore a textual one: protagonists of Mantegazza's *L'Anno 3000,* their point of reference is the *Year 3000* of the unnamed clinician-author. They are thus both readers of text and writers of it: their journey through the landscapes and cities of the year 3000 both constitutes the narration and drives it forward.[50]

With each stopover on their way to Andropolis, it becomes increasingly clear that at stake in their travels is Maria's education on her own role and that of institutions in the management of reproduction.[51] Maria's individual

50 For Pireddu, this *mise-en-abyme* functions as "a literary solution that highlights Mantegazza's proverbial self-centeredness," but also serves to "[enhance] the credibility of the future scenario" ("Introduction. Paolo Mantegazza, Fabulator of the Future" 24).

51 Despite a striking number of thematic parallels between Mantegazza's *L'Anno 3000* and Campanella's *La Città del Sole* (1602), Pireddu is among the few scholars who mentions the latter as one of several precursor texts ("Introduction. Paolo Mantegazza, Fabulator of the Future" 3). Capatti discusses another precursor, Albert Robida's *Voyage de fiancailles au XXe siècle* (1892). Instead of a novel that targets reproduction, Robida's *Voyage,* set in 1954, stages prenuptial travel as generative of matrimonial stability (12). Campanella, like Mantegazza after him, dedicates much of his text to the organization of sexual relations, particularly with regard to education, reproduction, and even hygiene (medical practice, alimentation, domesticity, etc.). The "Grand Magistrate of Generation, or the Grand Medical Doctor" oversees reproductive coupling (42). Men and women who have come of age (nineteen years for women; twenty-one for men) participate in a public display of unclothed athletic exercise, "displaying themselves nude in gymnastic exercises" (43). Their bodies are thus deemed fit or unfit for sexual reproduction, and couples are assigned based upon corporeal compatibility, "according to their respective corporeal proportions. [...] A large and beautiful woman is united with a robust and impassioned man" (43). On a designated night, at a designated hour (deemed by clinician and astrologer alike as favorable for insemination), and only after "the generators have washed themselves well," the couple is united for the reproductive act (43). In *3000,* Mantegazza shifts focus: instead of an examination of the generative couple, he places the product of their union on physiological trial. An examination of reproductive bodies seems inconsequential to the conferral of a license to reproduce (Paolo and Maria receive their permit without being subjected to such an examination).

reproductive role is, true to biopolitical imperatives, inscribed within a collectivity. In Foucault's words:

> [S]exuality represents the precise point where the disciplinary and the regulatory, the body and the population, are articulated. [...] Medicine is a power-knowledge that can be applied both to the body and the population, both the organism and biological processes, and it will therefore have both disciplinary effects and regulatory effects. (*Society Must Be Defended* 252)

The fictional year 3000 is situated where discipline and regulation intersect: the citizens of Andropolis are both self-governing and rigorously regulated. Disciplinary agents are replaced by civilians: "In the year 3000 there are no longer gendarmes, nor policemen, nor public security guards; every honest citizen is a gendarme, a policeman, and moreover also a judge" (*The Year 3000. A Dream* 101). Here is Bentham's Panopticon—"a generalizable model of functioning, a way of defining power relations in terms of the everyday life of men"—realized (Foucault, *Discipline and Punish* 205). The watchtower may or may not be empty, but subjects will behave as if it were constantly manned. In Foucault's now classic formulation:

> Power should be visible and unverifiable. Visible: the inmate will constantly have before his eyes the tall outline of the central tower from which he is spied upon. Unverifiable: the inmate must never know whether he is being looked upon at any one moment; but he must be sure that he may always be so. (201)

Or, as Thomas Flynn puts it in his analysis of vision in Foucault's account of modernity, "Disciplinary power, unlike traditional sovereign power, is exercised through its invisibility while imposing compulsory visibility on its subjects" (201). The disciplined future has already been realized in Andropolis, and therefore the role of the prison has been eclipsed by an even kinder, gentler institution—the school: "[T]his House is by no means a jail, as they were in earlier times, but a sort of school, where they correct the guilty, where they lovingly study the causes that can have led someone to commit a crime" (*The Year 3000. A Dream* 102). As I've mentioned, the educative project enabled by travel constitutes both the novel's content and its raison d'être. Yet Mantegazza saves a space in the year 3000 for Bentham's carceral Panopticon, albeit in a somewhat altered form: "The city of Andropolis boasts over fifty large theaters, which offer the widest range of shows to delight the eyes and ears—to delight the imagination and the heart. [...] Only the Panopticon, the largest and wealthiest of Andropolis's theaters, is the property of the State; its purpose in presenting its spectacles is to educate the people to appreciate beauty and the finer emotions" (155–156). It is almost as if Mantegazza anticipates Foucault's memorable antimetabole nearly a century later that "prisons resemble factories, schools, barracks, hospitals, which all resemble prisons," by adding to this list an artistic and/or cultural institution—the theater (Foucault, *Discipline and Punish* 228).[52] The Panopticon refigured as pedagogical (read: disciplinary) theater raises the question about the degree to which Mantegazza's text, itself structured by a pedagogical relationship, between Paolo and Maria, and between Mantegazza and his would-be readers, "stages" the encounter between the discipline and regulation of reproduction. It also brings up the question of visual perception, one which recurs throughout the novel (Paolo and Maria's travels are marked by statues and museums, which represent history visually: "you see before your eyes" is one of the novel's orienting refrains; futuristic medical technology is primarily scopic, that is, directed at objects of examination and fancy) and demarcates its ideological limits.[53]

Paolo's learned monologues on statecraft and reproduction both constitute and propel *L'Anno 3000*'s narrative. The limits of absurdity in the political utopias that Mantegazza presents as foils to Andropolis are often to be found in descriptions of how sexual relations are regulated there.[54] From a resident of The Land of Equality, Paolo and Maria learn: "A law passed in this very year requires all men to fecundate their wife only on the first of May. As

52 Here, Mantegazza seems also to anticipate Foucault's thinly veiled rejection of Guy Debord's well-known argument that consumer capitalism controls unknowing subjects through spectacle. Foucault writes, "We are much less Greeks than we believe. We are neither in an amphitheater, nor on the stage, but in the panoptic machine, invested by its effects of power, which we bring ourselves since we are part of its mechanism" (*Discipline and Punish* 217). In Mantegazza's novel, the Panopticon-as-theater dramatizes the very passage that Foucault will outline nearly a century later: from power as the spectacular display of the sovereign's credo "let live and make die," to power as both disciplinary and regulatory, modeled on the biopolitical injunction to "make live and let die."

53 Here, following Christian Metz and Martin Jay, I refer both to the Latin *scopium*, "to look at, examine," and to the noun *scope*, as in "something aimed at or desired." See: "scope, n. 2a," *The Oxford English Dictionary Online*; "-scope, *comb. form*," *The Oxford English Dictionary Online*; Jay; Metz.

54 For more on the political fantasies represented in *The Year 3000*, including that of European federalism, see Pireddu, "Introduction. Paolo Mantegazza, Fabulator of the Future" 24–26; 36–41.

for love, we make it at the same hour, every hour, when the special bell of the Government House tolls" (72). These lessons also prepare Maria, and readers, for arrival at Andropolis. There, the couple visits the Hygeia, Andropolis' hospital (or "Institute of Health"), named after the goddess of health and sanitation.[55] Aimed as it purportedly is at the celebration of the "fertile marriage" of the protagonists, the narrative foregrounds this visit: though the text seldom gestures forward, goings-on at the hospital are alluded to long before their arrival there. Yet, if these allusions are noted by readers, they fail to move the protagonists. As the couple interviews a citizen of Andropolis on the legal system, he mentions that recidivism happens only among "born delinquents who were *spared elimination* by errors in their brain exams" (emphasis mine; 103). Likewise, during their visit to the Health Ministry, the couple learns from a representative that, "Even in the year 3000 there continue to be people who are weak and destined to live short lives, and even though pathological newborns are destroyed, many imperfect organisms still remain who can neither find life at all pleasant nor make it useful to themselves and others" (111). Though the protagonists seem unmoved by the suggestion that living beings are "suppressed," and newborns "destroyed," the text thereby registers some anxiety about what awaits them at the Hygeia. Moreover, these exchanges forecast a disruption: at times, readers learn, "error" results in the survival of "imperfect organisms."

Paolo and Maria tour the grounds with the director of the *Hygeia*, who informs them about the professional hierarchy in place; the hygeians—those who practice the preventative medicine privileged by the hospital—are situated at the top of the hierarchy because they assess newborns "to verify whether they are fit for life" (135). After the couple visits the tuberculosis ward and the trauma wing, the anxiety of the text is displaced onto Maria; if, at earlier points in the narrative, Maria seems impervious to accounts of human "suppression," just before their arrival at the hospital, the text registers her heightened awareness. "'And now,' said the Director, 'let us go visit the division of the hygeians.'" Maria is apprehensive at the suggestion: "Maria, who had heard talk of the elimination of babies unfit for life but who knew nothing more about this, was rather anxious and unsure whether she should go into that department." Paolo, once again, steps in to manage

55 Between 1862 and 1881, Mantegazza published a journal first entitled *L'Igea. Giornale d'Igiene e Medicina preventiva*. He later renamed the journal to reflect a more popular agenda: *Il Medico di casa. L'Igea. Giornale d'Igiene e Medicina popolare* (Landucci 110).

her relationship to knowledge and desire. Critically, Maria's reproductive lesson hinges upon her (and by extension, the readers') *seeing* the horrifying medical intervention: "We must and want to see everything. Let's go." They enter a large waiting room full of wailing newborns and nervous mothers, who await "the doctor's sentence of life or death for their little children" (141). Mantegazza thus fuses the juridical language of the state ("life or death sentence") with that of medical hygiene, producing an immunitary model. Practitioners examine newborns with new technologies that render their skeletal, organ, muscular, and cerebral systems visible to clinical observers.[56] The narrator explains: "The hygeian took the baby, who was all but naked, and, stripping it completely, placed it on a sort of perch. Immediately a beam of light flooded over it, making it transparent, as though it were made of glass" (142).[57] From here, a diagnosis—physiological and psychological—is produced, and conveyed to a secretary, who records the diagnosis (as a "sentence," or "verdict"). In the event that the diagnosis is favorable, the examining doctors sign a certificate authorizing life, which is presented to the relieved mother: "Number 17: The child is healthy, sturdy, fit for life" (142). It is here that, to paraphrase Nicoletta Pireddu, the novel's celebration of scientific progress and technological innovation reveals its horrifying dystopian side ("Introduction. Paolo Mantegazza, Fabulator of the Future" 32).

56 Here, Mantegazza resembles David Horn's Lombroso, for whom: "the body was made an index of the interior states and dispositions of suspected individuals, a sign of the evolutionary status of groups, and a more or less reliable indicator of present and future risks to society" (*The Criminal Body* 1). Yet his mobilization of technology constitutes a departure from Lombroso, who made a name for himself by indexing immediately visible corporeal evidence of criminality. As Stewart-Steinberg writes succinctly: "Here the shape of an ear, of a chin, or the shifty glance of an eye already determine a subject in his or her future actions" (231).

57 This is the second time we read of the "perfected" X-ray, which makes the inner functioning of the human body visible to the observer. Just before their arrival at the Hygeia, Paolo and Maria encounter the following: "A delicate youth, pale and gaunt, was just then waiting to be called. The pneumopathologist asked him to remove his clothes and, once he was entirely naked, asked him to go stand in a sort of niche where suddenly the light that was illuminating the room disappeared, and all was plunged into darkness. Immediately after, though, the doctor directed a beam of light onto the naked man, who became as transparent as if he were made of glass. *One could see* the heart in its speeded-up, irregular beating, and *see* the lungs dilating and contracting rhythmically, and *see* all the viscera of the belly, as if that man had been opened by an anatomic knife, one could even make out the marrow in the depths of the bones" (emphasis mine; 137).

After witnessing several similar "verdicts," once again, Maria's anxiety partially prepares readers for what is to come: the medically assisted execution of infants. Yet what reads initially as an entirely regulatory system of state medicine is complicated in the final instance, when Mantegazza introduces the caveat of maternal consent:

> Maria sperava che le visite avrebbero avuto tutte un analogo risultato, per cui non avrebbe assistito alla distruzione di nessuna creatura, ma ecco che il numero 20, un bambino gracilissimo e che per di più era nato di otto mesi, sottoposto all'esame dell'Igeo fece aggrottare le sopracciglia al medico. (116)
>
> [Maria hoped that all the visits would have a similar outcome, so that she would not have to witness the destruction of any creature, yet then came number 20, a very frail baby who, in addition, had been born at eight months and who, on being submitted to the examination of the hygeian, brought furrows to the brow of the doctor] (143)

The clinician thus diagnoses the newborn: "Baby very frail, tubercular, unfit for life."

Yet just as the authoritative syntax of diagnosis becomes indistinguishable from an institutionally-mandated death sentence, the narrative takes a curious turn: the clinician confronts the mother with, "And so?" Weeping, she replies, "Yes." The narrator interjects, euphemistically translating the words of the clinician: "That 'And so?' meant: So, do you allow your baby to be eliminated?" With the consent of his mother, the whimpering baby is thrown into an incinerator:

> E infatti, un inserviente prese il bambino, apri un usciuolo nero, posto nella parete della sala e ve lo mise, chiudendo la porticina. Fece scattare una molla, si udì un gemito accompagnato da un piccolo scoppio. Il bambino innondato da una vampa di aria calda a 2000 gradi era scomparso e di lui non rimaneva che un pizzico di ceneri. (117)
>
> [And in fact an attendant took the baby, opened a small black portal in the wall of the room, and put it in there, closing a small door. A spring was released, a cry was heard, accompanied by a little explosion. The baby, enveloped by a flare of hot, 2,000-degree air, had disappeared, and only a bit of ash remained.] (143)

Here is where discipline and regulation meet: institutional agents (in this case, clinicians) produce a "sentence," and mothers are to act as knowing subjects, educated in the tenets of institutional hygiene that conducts them, as compliant citizens of Andropolis, to authorize the execution of their infants. Fredric Jameson points out how moves toward the totalizing institutionalization of life constitute a recurrent paradox within utopian narrative:

> Indeed, one of the basic constraints of the form would seem to be the incompatibility within it between action or events and that timeless map-like extension of the non-place itself: in other words, if things can really happen in Utopia, if real disorder, change, transgression, novelty, in brief if history is possible at all, then we begin to doubt whether it can really be a Utopia after all, and its institutions—from a promise of the fulfillment of collective living—slowly began to turn around into their opposite, a more properly dystopian repression of the unique existential experience of individual lives. ("Of Islands and Trenches" 17)

Vittorio Roda makes a similar observation about *3000*, calling it a "monument to nineteenth-century optimism" while asserting that:

> [T]here's something troubling in the air: the undesired and unanticipated flipside of that very optimism, the shady zone tucked into the folds of an 'elsewhere' that the implacably normalizing rationalism of Mantegazza renders [...], in spite of the expressed intentions of the author, perilously close to the ruthless universe of the dystopian. (148)

The scene at the Hygeia overwhelms Maria, who pleads with Paolo to move on from this "chamber of horrors," yet Paolo insists that they watch one last examination. This final encounter introduces the possibility of a mother's refutation of the clinician's diagnosis and sentence. When an attendant confronts the mother of an infant whose outwardly delicate constitution recalls that of the baby who had met with the unhappy fate of incineration, the mother rejects his diagnosis of "Baby with serious heart weakness, unfit for life" (144). As he menaces her with, "And so?" she appropriates his language, wielding it in the service of her own refusal: "And so? And so? So, no!" This declaration marks her exit: she hurriedly escapes from the hospital, clutching her fragile infant. Critically, no guard, no policeman intervenes to stop her. She is "free" to leave, with only her conscience left to answer to: "Poor woman!" the hygeian laments, "How often she will regret that *no*. She thinks she's being a good mother, instead she's merely being a cruel one" (145). The inherent cruelty of the state's eugenic murder of infants is thus displaced onto an individual mother's choice to defend the life of her ailing baby. As an agent of the state and the so-called public good, the hygeian's accusation of cruelty is of course waged at this individual as a kind of social or collective mother, a would-be mother of the 'hygienic' nation. He evokes the logic of individual sacrifice for the common good, which in Mantegazza's biopolitical utopia is understood as necessitating an immunitary defense.

What are the narrative consequences of this discovery? The novel is unable to arrive at a celebration of reproductive bliss among its protagonists—no happy, robust babies are born to Paolo and Maria—because the project of Maria's sexual education is thwarted by this gesture of female resistance to the disciplinary and regulatory mechanisms of the Hygeia. Furthermore, it is at this critical point, when the text exposes the destructive mechanisms at work in the eugenic utopia that it has labored to construct, that the logic of the text takes a sharp turn. If the locus of the utopian fantasy has been the possibility of disciplining and regulating healthy life (in an immunitary key: by exposing pathological elements to death), this moment of refusal constitutes a disruption of that fantasy. Like immunitary logic itself, the very notion of *3000* as a eugenic and/or 'hygienic' utopian novel contains troubling traces of its opposite: a dystopian future in which human babies are tossed into incinerators, presumably making life unbearable for their parents. At a narrative level, the mother's imperviousness to hygienic discipline and regulation result in the survival of the "imperfect organisms" anxiously evoked at various points of Paolo and Maria's journey. The result is that, instead of figuring Maria's hygienically enlightened parturition, the novel concludes by figuring that of Paolo. The "birth" that is ultimately celebrated (Paolo's psychoscope) is the last in a constellation of Mantegazzian inventions aimed at further rendering bodies "transparent."

The technological innovations that Mantegazza envisions enable the eugenic project of the hygeians, rendering physiological and psychological man legible and intelligible. Perhaps not surprisingly, despite the fact that national origins, like national languages, are staged as relics of the past, it is Italian genius that propels these technological developments:

> Quest'altra statua è innalzata al dottor Micali, medico italiano, che nel secolo XXV, perfezionando la luce Rontgen riuscì a render trasparente tutto quanto il corpo umano, permettendo cosi di vedere ad occhio nudo

> il cervello, i polmoni, il cuore; tutti quanti i visceri e perfino il midollo delle ossa. (106–107)
>
> [This other statue is erected to Doctor Micali, an Italian physician who, in the twenty-fifth century, by perfecting Röntgen light, managed to make the whole human body transparent, thereby allowing us to see with the naked eye the brain, the lungs, the heart, all the inner organs, and even the marrow of bones.] (134)

Newborns are examined and (almost always) exterminated if their bodies are (in)visibly marked as defective. In order for the body to harbor traces of moral or biological atrophy, technology must be mobilized in order to assist the human eye; physiognomic and/or chromatic race is thus surpassed by medico-scientific innovation, which renders the skin "as transparent as [...] glass" (136). The scientific "birth" that marks the close of the novel complements this physiological visibility with psychological transparency: Paolo's invention of the psychoscope, "a little instrument the size of pocket binoculars [...] that allows us to easily read the thoughts of others to whom it is directed," is awarded the Cosmic Prize by the Academy of Andropolis (189). Instead of a public ceremony celebrating the protagonists' fertile marriage, then, the public ceremony staged at the novel's conclusion is in celebration of this invention aimed at reducing bodies to legible text.

These inventions serve to render physiognomic and epidermal race superfluous: the truth of the body is located instead within the microscopics of the brain, organs, and muscular systems. For Mantegazza, this truth is accessible thanks to the work of technology, understood as both the text itself and the medical apparatuses it produces. If, on the one hand, this logic undermines the potency of visible race (skin color, physiognomy, etc.), on the other hand, it leaves intact that of invisible race (blood, biological essence). Truths of psychological and physiological systems are no longer legible on the skin or on the face; the text and other technologies thus intervene with inventions aimed at exposing what lies within. Yet the novel does not arrive unproblematically at this conclusion, wherein bodies are reduced to little more than intelligible and predictable text. Indeed, the thematic conclusion might be said to be dislocated from the ideological conclusion offered by the novel. If the novel "ends" here, with Paolo's technological triumph over psychological opacity, where to situate the narrator's account of the biological surpassing of racialized variation? Why is the road to the full disclosure of biological truth marked with a fantasy about the historic dissolution of race (ostensibly having occurred long before the present tense of the text)? In spite of itself, the novel seems to know that it is not beyond modes of racialization that posit the immediate visibility of race. Indeed, if it were, why would a central thematic focus of the novel be the protagonists' arrival at the eugenic laboratory of the Hygeia? If racialized variation has already given way to Darwinist selection and regeneration ("a new type, indefinitely cosmopolitan") why are the text's technological innovations aimed directly at uncovering the very biological truths that race either discloses or causes?

Returning, then, to Paolo and Maria's visit to the Hygeia: it concludes, as I have noted, with the refusal of a mother to sacrifice her child to the mandate of public hygiene. Soon thereafter, the couple visits the Museum of Andropolis, which contains artifacts that recount—true to Mantegazzian form—the entire physiological and psychological history of man, arranged for at-a-glance consumption.[58] Within the museum are arranged plastic replicas of human beings:

> [*T*]*i vedi davanti agli occhi* l'Adamo selvaggio e irsuto dell'epoca quaternaria, l'uomo delle caverne, l'uomo neolitico e infine tutta la lunga schiera di razze più moderne e che sono però già scomparse dalla superficie della terra; quali gli Australiani, i Maori, gli Ottentotti, i Boschimani, molti Negri, i Guarani e tante e tante altre razze, di cui per alcune però rimangon le traccie nei contemporanei del secolo XXXI. (145)

> [*You see before your eyes* the wild, hirsute Adam of the quaternary age, the caveman, Neolithic man, and finally the whole long array of more modern races that have however already disappeared from the face of the earth—such as the Australians, the Maori, the Hottentots, the bushmen, many blacks, the Guarani, and so many other races, some of whom, however, remain, albeit only in traces, in the thirty-first century.] (emphasis mine; 169)

58 The museum thematizes Mantegazza's oeuvre as a whole, aimed, as I have mentioned, at narrating "the entire history of the human family." It is tempting to forge connections between the circular layout of this fictionalized museum in Andropolis and Mantegazza's own in Florence. Such work would entail reconstructing the organization of the Museum of Anthropology and Ethnology as it stood in 1897. Though, today, Mantegazza's Museo follows a similarly circular layout, it has only been housed at its current site since 1910.

Epidermal and physiognomic race reemerges in the space of the museum, where racialized man is showcased for viewers. Despite the fact that these visually marked bodies are staged as relics of the past, they signal a kink in the Mantegazzian logic that I have been tracing. In order for the text to suggest that it has surpassed physiognomic and/or epidermal race, it must stage its dissolution; populating its futuristic world are not physiognomically or epidermally marked racial others, but instead the "indefinitely cosmopolitan," and ostensibly post-racial, type. In Mantegazza's post-racial future, the human eye alone is unable to read racial difference and/or signs of physiological inferiority; technologically assisted forms of viewing are required to penetrate the skin and evince physical defects that lie beneath the skin. Ultimately, however, the text falls back upon an affirmation of the very types of seeing that it renders obsolete.[59] How will viewers in the year 3000 consume these racialized artifacts as such without resorting to the very modes of physiognomic and/or epidermal racialization (or what Sekula deems physiognomy's "everyday nonspecialist empiricism" (11)) that the text has attempted to surpass? It is precisely here that the text turns back upon itself. Just as it becomes clear that year 3000 museum patrons require the modes of seeing that the fantasy of the text has labored to dissolve, the novel ushers in a self-critique that exposes the utopian project for what it is: a product of the "bizarre fantasy" of a nineteenth-century thinker.

From the plastic replicas of racialized man, Paolo and Maria continue on to the naturalist wing of the museum, which houses the extraterrestrial exhibit: "[S]ome naturalists, richer in imagination than in science, imagined how the planetary inhabitants might appear and rendered them in drawings or sculpture" (170). Paolo and Maria giggle as they observe the naturalists' representations of life forms on other planets:

> O Maria Mia, come son buffi questi angeli planetarii, come sono grotteschi, sopratutto come sono impossibili! Mi par che i naturalisti, che li hanno scoperti, dovevano conoscere ben poco l'anatomia comparata e ancor meno la biologia. [...] Guarda qui, questo abitante di Venere quanto è buffo! Gli hanno appiccicato due ali e questo è il

59 Here I refer to Jonathan Crary's argument that a new kind of seeing subject emerged in the nineteenth century, whose capacity for vision lay in his physiological composition rather than in the exterior mechanics of optics. Critically, this new subjective vision—articulated most forcefully by Goethe and Schopenhaur—"coincided with the making of the observer into a subject of new knowledge and new techniques of power" (79).

> sogno più antico, che ha creato gli angeli delle teogonie cristiane, delle maomettane, e di tante altri religioni. L'uomo ha sempre desiderato di poter volare e attaccandosi due grandi ali di oca, di cigno o di aquila, ha fabbricato i suoi angeli. (147)
>
> [Oh, my dear Maria, how comical these planetary angels are, how grotesque, above all, how impossible! It seems to me that the naturalists who have discovered them must have known very little comparative anatomy and even less biology. [...] Look here, at this 'Venusian!' How funny it is! They've stuck two wings on it—which is the oldest dream we have—and created the angels of theogonies of the Christian, Islamic, and a good many other religions. Man has always longed to fly, and by tacking on two big goose or swan or eagle wings he has created his angels.] (171)

The inherent limits of the naturalists' historically contingent worldviews are represented by their Venusians, who are little more than mannequins plastered with goose feathers. The narrator continues with a similar critique:

> In tutti quei mostri, però non si poteva trovare un solo organo che già non esistesse nell'uomo o in altri animali, per cui la nuova creatura planetaria non era che un mosaico di membra diverse prese ora agli uccelli, ora ai pesci, o agli insetti o ai molluschi. (148)
>
> [In all these monsters, however, one could not find a single organ that did not already exist in man or in other animals, so that the new planetary creature was merely a mosaic of different body parts taken now from birds, now from fish or from insects and mollusks.] (171)

This passage reads as self-referential; it suggests that there are ideological boundaries in place that serve to limit the work of futuristic projection. In this case, the targets of the critique are both the naturalists—who depict extraterrestrial creatures with pencils, plastic, and goose feathers—and the novel itself, which fantasizes about the end of race as it mobilizes technologies of racial visualization in order to do so. This self-ironization underscores the text's acknowledgement of its own boundaries, marking what I consider the ideological resolution of the novel: in spite of itself, the text recognizes that it is not beyond the visual logics of racialization that it has toiled to overcome.

That the novel's ideological parameters should be exposed by how it approaches the apparent legibility of bodies brings up some of the questions with which this chapter began. Part of what critics have found so simultaneously compelling and threatening about Mantegazza's work is how it brings sexualized and racialized bodies into the realm of visibility through writing and Mantegazza's use of photography, which he claimed would facilitate even more readily a "comparative reading" of human bodies. Likewise, Mantegazza's feverish publication of volumes upon volumes of popular prose, which he himself refers to as textual offspring, must be read as therapeutic, as an attempt to heal the Italian masses, delivering them from what would otherwise be their biopolitical decline. Subtending such a rhetoric of decline is the ideological fantasy of the racialized Italian national body. Like Maria traveling through the institutions of Andropolis, readers are to learn the joys of personal and collective hygiene, and all of the "freedoms"—from the oppressive regulation of Tyrannopolis, and even the Land of Equality; from pain, sickness, and premature death—that it enables.[60] And at the apex of this biopolitical project lies nothing short of an immunitary imperative. Indeed,

60 "Human life too has been considerably prolonged, thanks to the rising prosperity of the poor classes and all the progress in hygiene. Whereas in the nineteenth-century an average life span was between twenty-eight and thirty-six years, today the average planetary life span is seventy-two years and in some healthier regions up to eighty-five years" (112). Pireddu lists these and many other of the "human and social benefits" produced in the novel ("Introduction. Paolo Mantegazza, Fabulator of the Future" 27). For Žižek, following Lacan, this is how Marx invented the symptom: "The 'symptom' is, strictly speaking, a particular element which subverts its own universal foundation, a species subverting its own genus. In this sense, we can say that the elementary Marxian procedure of 'criticism of ideology' is already 'symptomatic': it consists in detecting a point of breakdown *heterogeneous* to a given ideological field and at the same time *necessary* for that field to achieve its closure, its accomplished form. This procedure thus implies a certain logic of exception: every ideological Universal—for example freedom, equality—is 'false' in so far as it necessarily includes a specific case which breaks its unity, lays open its falsity. Freedom, for example: a universal notion comprising a number of species (freedom of speech and press, freedom of consciousness, freedom of commerce, political freedom, and so on) but also, by means of a structural necessity, a specific freedom (that of the worker to sell freely his own labour on the market) which subverts this universal notion. That is to say, this freedom is the very opposite of effective freedom: by selling his labour 'freely,' the worker *loses* his freedom—the real content of this free act of sale is the worker's enslavement to capital. The crucial point is, of course, that it is precisely this paradoxical freedom, the form of its opposite, which closes the circle of 'bourgeois freedoms'" (*The Sublime Object of Ideology* 22).

I read Mantegazza's modernity in *L'Anno 3000* not, as others have, in his prophetic poetics—he "predicts" the outbreak of World War I, the demise of the socialist experiment, the formation of the European Union, and he anticipates the invention of the CAT scan or, even more recently, the U.S. Transportation Security Administration-enforced body scan, and "clean" energy, nor in his many other fanciful futuristic inventions such as popular air travel à la Jettsons and the Pantomass, a personal massage contraption that conforms to individual bodies like a well-tailored suit, and so on.[61] Instead, taking my cue from Roberto Esposito's contention that immunization—the hermeneutic that "lays bare the lethal paradox that pushes the protection of life over into its potential negation"—is what defines modern political subjectivity, what renders this novel 'modern' is how Mantegazza stages the encounter between disciplinary and regulatory regimes of life and death (Esposito, *Bíos* 116).[62] Indeed, it is only in this biopolitical key that one can interpret Mantegazza's relationship to technology. In addition to the medical technologies I have already discussed, the year 3000 is home to new networks of global communication modeled on the human body:

> E così come nel nostro corpo ogni organo, ogni cellula ha la propria vita indipendente e solo si mantiene collegato nella grande federazione e nella grande unità dell'organismo per mezzo del sistema nervoso e del sistema sanguino; così nel nostro pianeta ogni Comune vive da sé, a per mezzo dei fili telegrafici che rappresentano i nervi, comunica con Andropoli, che è in una volta sola cervello e cuore del gigantesco organismo planetario. Fra i Comuni e il centro vivono poi tanti centri minori che sono le *Regioni*, le quali rappresentano i gangli. (emphasis in original; 79)

> [And just as with our body every organ, every cell, has its own independent life and only maintains itself in a linkage with the great federation and the great unity of the organism by means of the nervous and blood systems, so on our planet every commune lives by itself, but through the telegraphic wires that represent the nerves it communicates with Andropolis, which is at once the brain and the heart of the giant planetary organism. Between the communes and the center, then, dwell many

61 Pireddu discusses these and other Mantegazza predictions in *L'Anno 3000* "Introduction. Paolo Mantegazza, Fabulator of the Future" 34–42.

62 Esposito explains, "[It is] the modality of immunity through which the Modern thinks the figure of the subject" (56).

minor centers, which are the *regions*, representing the ganglia.] (emphasis in original; 108)[63]

The planetary community of 3000 is connected through information technologies modeled on the human body, a body that has been rendered "as transparent as glass" thanks to still other medical technologies aimed at disclosing corporeal defects that lie beneath the skin. And yet this very transparency conceals an immunitary imperative that forecloses the very community it claims to preserve. Mantegazza's technologies of visualization might also be considered in light of the ideological fantasy of the racialized Italian body, insofar as they respond to an anxious question that the novel suggests to vexed thinkers like Mantegazza: what if, upon closer and ever-more technologically enhanced inspection (be it the photograph or the viewing machine that renders bodies "as transparent as glass"), the traces of Italian racial identity are nowhere to be found? What if the body fails to register the racial truth of Italian identity?

This chapter has been concerned with how Mantegazza produces the racialized body through immunizing technologies of visibility. The next chapter follows the flow of race beneath the skin in Gabriele D'Annunzio's novels the *Romanzi della rosa* and his Fiuman discourses. At Fiume, technology is mobilized not, as in *L'Anno 3000,* to evaluate biological fitness, but instead to propel the racialized body skyward. The passage from Mantegazza's *L'Anno 3000*—a mediocre and poorly received novel that represents what critics figure as Mantegazza's amateurish engagement with literature to thinly veiled scientific ends—to an imposing figure like D'Annunzio, whose language and style are instead heralded as the pinnacle of liberal Italy's literary innovation, requires some comment. Though avowed foes in life, Mantegazza's version of Darwinism influenced D'Annunzio's composition of his first best-selling novel, *Il Piacere* of 1889 (Pireddu, *Antropologi alla corte della bellezza*). If Mantegazza's oeuvre can be considered as at least part of the raw material from which D'Annunzio—a notorious gatherer of facts and lifter of text—drew in composing his literary masterworks, an analysis of race and (re) productivity within these two bodies of work presents interesting interpretive

63 At the risk of celebrating yet another Mantegazzian "prophecy," Lisa Cartwright and Brian Goldfarb note: "After World War II, discourses of neurology and machine computing became mutually constitutive. Descriptions of the nervous system as a model for machine intelligence were paralleled by characterizations of the nervous system as a conductive mediating network" (127).

possibilities. That is, if the "scientific" content—rather than the literary sophistication—of Mantegazza's writings was digestible to D'Annunzio, following the thread of race and (re)productivity from one set of writings to the next allows readers to observe yet another passage: from the laboratory of the scientist, to the inept hands of the literary dilettante, in D'Annunzio's texts, readers may behold the refashioning of these concepts in the studio of a master. D'Annunzio tethers rhetorics of race and (re)productivity to a variety of other formulations about Italian modernity and preeminence. Race is inscribed within a poetic constellation that figures blood and territory as the rhetorical conditions for Italy's conquest of modernity. Despite the migration of these themes across uncharted stylistic terrain in D'Annunzio's oeuvre, their ideological coherence nevertheless remains bound to their first and clumsy enunciation in *fin-de-siècle* social science.

CHAPTER THREE

Mutilated Limbs

Post-Unification cultural production depicts nation building as a kind of ideological work (political, literary) that produces hard-working national subjects. In both Leopoldo Franchetti's colonial biopolitics and Paolo Mantegazza's immunitary public health discourse, the production of Italians as *vital subjects* requires figures of labor productivity and racialized biological reproduction. Massimo D'Azeglio's famously ill- but nevertheless oft-cited dictum "We have made Italy, now we must make Italians" ("Si è fatta l'Italia, ma non si fanno gl'Italiani") might thus be rethought in terms of how this production or 'making' (with an emphasis on the verb "fare") is contingent upon other productivities: labor, biological, but also the discursive or aesthetic productivity of certain racial logics within the languages of liberal nationalism. *Fin-de-siècle* Italy's literary and political *provocateur* Gabriele D'Annunzio's early novelistic trilogy the *Romanzi della rosa* (1889–1894) further illustrates how sexual reproduction—and attendant figures not only of fecundity and prolificacy, but also female sterility and hereditary degeneracy—is a particularly dense site of rhetorical articulation in racial discourse. Critical to *making* Italians (*fare gl'italiani*), well before Mussolini's demographic policy and his more explicitly racist and imperialist projects, is a reproductive and racializing imperative, encapsulated by the admittedly much older expression *fare razza*, literally, "to make race," or, in other words, "to reproduce."

That biopolitics forges a relationship between the reproduction of the biological individual and the political collectivity has consequences for D'Annunzio's literary and political production—from the *Romanzi della rosa* to his Fiuman discourses (1919–1920)—given his avowed commitment in (not only) these works to hereditary genius, literary and biological genealogy, political spectacle, and/or territorial expansion expressed through

Fig. 3.1 Souvenir "D'Annunzian tissues," featuring an image of the Ansaldo SVA biplane D'Annunzio used in the *Volo su Vienna* (1918), on display at the Vittoriale (Photo by author.)

physiological rhetoric.[1] Reading the thematic trajectory of his physio-psychological *Romanzi della rosa,* it consists of three narratives that articulate a version of this relationship between subject and collectivity. Tracing the novels schematically reveals a sexually potent but non-reproductive male protagonist as the last representative of a family line (*Il Piacere,* 1889); a father threatened by biological and genealogical "intrusion" in the form of an adulterine fetus (*L'Innocente,* 1892); and a male protagonist who, plagued by monstrous heredity (from both "fathers," from his sterile female partner, and from the biological community of the Abruzzi countryside, in *Trionfo della morte,* 1894), is forced to definitively remove himself (and his lover) from the genetic scene through homicide and suicide. The generative thrust of each novel thus expands progressively in focus from individual, to genealogy, to racialized collectivity.[2] This passage from the individual to the collective is enacted

1 Lucia Re proposes a genealogy of D'Annunzio's interest in *stirpe, razza,* and *sangue.* Re's reading privileges the fifth section of D'Annunzio's *Il trionfo della morte* and *Le vergini delle rocce* as significant sites for the emergence of D'Annunzio's explicitly racial discourse ("Italians and the Invention of Race" 12–13). By drawing the focus to the role of blood in D'Annunzio's sexual and reproductive discourses of the earlier novels of the *Romanzi della rosa,* I propose an alternative path.

2 In English, the titles are translated as The Child of Pleasure, The Victim, and The Triumph of Death.

through a transfer of the rhetorical significance of *sangue* from *Il Piacere* to *Trionfo della morte*; while in the first blood remains bound to an erotic, yet not procreative, discourse, by the final installment, blood stands in for racialized descent. A reading of D'Annunzio's speeches and writings at Fiume in the light of D'Annunzio's earlier racial discourse illustrates how D'Annunzio recasts this relationship between blood and race at Fiume, producing a biopolitical model according to which the blood of mutilated soldier-patriots is mobilized not to signal genetic decline, but instead to ensure the rebirth of an Italian race.

A reading of the *Romanzi della rosa* alongside D'Annunzio's Fiuman texts is authorized at the thematic level as these early D'Annunzian novels forge relationships between blood and (re)generation (romantic, spiritual, intellectual, and/or biological) that anticipate the emergence of a naturalized population of Italians in the rhetorical terrain of Fiume. As much scholarship on the trilogy contends, these novels, more than D'Annunzio's earlier prose works (*Terra vergine, Novelle della Pescara*), contain sustained meditations on the crafting of male subjectivity (be it that of hero, anti-hero, or proto-*superuomo*), particularly with regard to how this process is structured by (hetero)sexual relations.[3] Renato Barilli suggests that the destitution of the characters upon whom D'Annunzio focuses his earlier prose inhibits him from developing a model of modern subjectivity, and that it is for this reason that his landscape shifts from the bleak lands of the Abruzzi to the opulent drawing rooms of Rome in *Il Piacere*:

> The problem of psychological order that interested him could not be confronted by dragging along the enormous economic and environmental handicaps that weighed on his characters taken from "the fourth state." He needed to sound it against human figures who were sheltered from the grip of neediness, and therefore free to dedicate themselves calmly to healing the "soul." (40)

3 For a discussion of this imposition of hyper-heterosexuality on D'Annunzio and his work through a reading of relationships between men in *Il Piacere, L'Innocente* and *Forse che sì forse che no*, see Duncan, *Reading and Writing Italian Homosexuality* 17–41. For a biographical account of the writing and publication of the trilogy (including an examination of D'Annunzio's epistolary exchanges, as well as of his French and German sources), see editor Annamaria Andreoli's introductions to *Il Piacere* (1105–1139), *L'Innocente* (1240–1251), and *Trionfo della morte* (1260–1297) in D'Annunzio, *Prose di romanzi*. For a study of D'Annunzio's novels that focuses upon the formulation of male subjectivity therein, see Goudet. For a call for the revaluation of D'Annunzio's prose, see Barilli.

I suggest that these "handicaps" crop up again in different guises, both in his trilogy and at Fiume: while the link between the peasants of *Terra vergine* and the *Novelle della Pescara* is drawn most explicitly in the popular religious scenes depicted in *Trionfo della morte*, their figuring in this final novel of the trilogy relies upon an imbrication of blood and heredity that is rehearsed in the protagonists—Andrea Sperelli (*Il Piacere*) and Tullio Hermil (*L'Innocente*)—of the novels that precede it. The "arrival" of the trilogy in *Trionfo della morte* at an intersection of blood and race through the flawed bodies of those who populate the text enacts a re-ordering of the ideologemes of blood, sex, race, and reproduction that are at play throughout the trilogy, setting the stage both for Claudio Cantelmo's "failed" reproductive charge in *Le vergini delle rocce* (1896) and, I argue, for the sacralization of blood through the mutilated body of the soldier-patriot at Fiume.[4]

The novels address the development of male subjectivity through licit and illicit heterosexual relationships (Andrea with Elena and Donna Maria; Tullio with wife Giuliana and lover Teresa; and Giorgio with Ippolita), and through their related concerns with heredity, they illustrate the inextricability of race from the production of sexuality in a way that anticipates Cantelmo's famed Nietzschean assignment in *Le vergini delle rocce* to, "preserve the ideal riches of [his] race [*stirpe*] and [his] own conquests in a son" (D'Annunzio, *The Maidens of the Rocks* 62).[5] Jacques Goudet calls D'Annunzio's novelistic project as a whole a search for "the essence of man and of the I" and he contends that this quest inevitably intersects with "the biological notion of heredity that the science of the time had taken as its analytical object" (11). Goudet continues, reading D'Annunzio's deployment of *razza* and *stirpe* as a necessary consequence to this encounter between *fin-de-siècle* ontology and genetics: "Essential being could not die with the individual. Effectively, D'Annunzio senses that the *I* receives and transmits it. As a consequence, notions of race

4 In her reading of the *Le vergini delle rocce* as, far from a novel about procreation, a novel about the fathering of a father, Barbara Spackman points out, "For much criticism of the novel [...] the fact that no child is conceived, let alone born, is oddly elided" (*Fascist Virilities* 175, n. 57).

5 Annamaria Andreoli locates the source for this assignment in Nietzsche's *Thus Spoke Zarathustra*. Interestingly, the passage she cites as a source does not advocate procreation in the name of the preservation of a *stirpe*, but in the name of the valorization of individual victory and freedom (1136). Alternatively, in her analysis of D'Annunzio's understanding of art as a kind of gift giving that defies the bourgeois-materialist logic of utilitarian exchange, Nicoletta Pireddu reads Cantelmo's charge in terms of its Nietzschean intertext *The Will to Power* ("Gabriele D'Annunzio").

[*razza* and *stirpe*] inevitably came up in his thought" (11). Like the confrontation of Paolo Mantegazza's physiology with "typical nineteenth-century" racial science, here race is figured as the "inevitable" outcome of D'Annunzio's philosophical inquiry. The *Romanzi della rosa* contain a variety of concerns with heredity, both retrospective and prospective. Yet D'Annunzio did not merely gather ready-made notions of *razza* and *stirpe* in order to place them, integral and unchallenged, on novelistic display; instead, these novels are sites within which D'Annunzio stages a negotiation of these racial terms. As Lucia Re puts it, "For D'Annunzio in the 1890's, and even to some extent in later decades, 'razza' and 'stirpe' are hardly fixed biological (or even social or cultural categories), but rather shifting rhetorical constructs, to be deployed in different ways as circumstances and his own image-making process dictate" ("Italians and the Invention of Race" 15).

It might be tempting to line up a distinction between *stirpe* and *razza* with what Foucault has argued was a shift from an aristocratic "symbolics of blood" to a bourgeois "analytics of sexuality." He writes:

> [The bourgeoisie] must be seen [...] as being occupied, from the mid-eighteenth century on, with creating its own sexuality and forming a specific body based on it, a "class" body with its health, hygiene, descent and race: the autosexualization of its body, the incarnation of sex in its body, the endogamy of sex and the body. There were doubtless many reasons for this. First of all, there was a transposition into different forms of the methods employed by the nobility for marking and maintaining its caste distinction; for the aristocracy had also asserted the special character of its body, but this was in the form of *blood,* that is, in the form of the antiquity of its ancestry and the value of its alliances; the bourgeoisie on the contrary looked to its progeny and the health of its organism when it laid claim to a specific body. The bourgeoisie's "blood" was its sex. (*The History of Sexuality* 124)

Indeed, common sense understandings of the distinction between *stirpe* and *razza* might reproduce such a distinction: *stirpe* as illustrious heredity or stock, and *razza* as a somehow distinctively more "bourgeois" or "modern" race (all attendant fascist associations therein). Yet such a distinction does not hold, for the simple fact that both *stirpe* and *razza* are such mobile signifiers that, throughout the Italian literary tradition, they are aligned with either bloodline or progeny at any given moment. Aligning *razza* with reproduction instead of retrospection and *stirpe* with genealogy instead of generation is thus too facile a distinction. Early modern authors, from Dante to Tasso, Giovanni Gherardi (1367–1446) to Matteo Maria Boiardo (1441–1494) deployed either *razza* or *stirpe* in order to make a claim about illustrious heredity, past or future. Each posited a relationship between genealogy and (regional) territory (Gherardi wrote, "Disceso son di stirpe fiorentina/ del Garbo nato e nobil cittadino"), which would later be bound to blood and nation as nationalism was more explicitly racialized.

Between the eighteenth and nineteenth centuries, there was a metaphorical explosion of *razza*, while *stirpe* maintained its "roots" in genealogy, retrospective or prospective. What becomes clear in an analysis of the two terms is that the most mobile of the two signifiers is *razza*, for the multiple discourses to which it pertains include zoology and biology, but also sex ("fare razza"), class ("perdersi la razza"), and nation ("razza italiana"). The Italian *razza* has indeed undergone a "colorful" or "diverse" metaphorization: *razza* is used to describe anything from breeding to manufacturing, from humans and horses to wine and violins. In the eighteenth and nineteenth centuries—preceding the processes by which so-called scientific race was solidified as a discourse through its alliance with modern psychoanalysis—*razza* was unhinged, however tentatively, from its traditional bases in zoology and biology, as it became a powerful signifier for a more generalized multiplicity and its ordering through classification. For example, the Italian expression "farne di tutte le razze" ("to do all kinds of things," *lit.* "all races") simultaneously posits a potentially disorienting disorder ("all kinds of things") and its abatement through an imposition of typology (the things are contained as *razze*). A similar example is, "vederne di ogni razza"—to experience a great variety of things (races). As *razza* was distanced from its direct relation to the body, into the generic realm of the (even inanimate) "type," it nevertheless remained tethered to a rhetoric of value: "di razza" (and variations there upon) is used to connote worth.

That D'Annunzio uses the *Romanzi della rosa* as sites of negotiation of the signifiers *razza* and *stirpe* has consequences for D'Annunzio's later literary and political projects. The trajectory of genetic decline enacted by these earlier novels is recast at Fiume as, far from a deficit, necessary to the exercise of biopolitical and territorial dominion. The *corpo mutilato* of the Fiuman soldier-hero is a figure for Italy's *vittoria mutilata* in World War I.[6] This flawed

6 For a detailed history of Italy's participation at the Paris Peace Conference, see Burgwyn. The history focuses on diplomatic relations rather than the production and diffusion of the "legend," and includes anecdotal information on D'Annunzio's role in

embodiment represents not defeat or resignation, but instead the condition of possibility for D'Annunzio's restoration of Italian primacy through aerial conquest.

The Eroto-Poetics of Blood in *Il Piacere*

A number of scholars have explored D'Annunzio's decadent poetics by noting that the "monstrous" hero/protagonist of *Il Piacere,* Andrea Sperelli, is figured as the last in a noble line.[7] However, the fact that in order to figure Sperelli as such D'Annunzio avails himself of two terms—both *razza* and *stirpe*—has received less attention. Perhaps this is due to the fact that their synonymic relationship renders the terms less than noteworthy. But can this shift be chalked up to a preference, whimsical or recherché, for *varatio*? While such an allegation might appease critics who emphasize his "decadent" penchant for arbitrary lexical artistry (or pastiche), given the complex role that an oscillation between the terms has played throughout the Italian tradition (and particularly within fascist rhetoric, to which D'Annunzio has been positioned, however problematically, as a precursor), the formulation in question requires some unpacking.[8] After an extended elaboration of five centuries of illustrious Sperelli, the oft-cited passage reads:

> Il conte Andrea Sperelli-Fieschi d'Ugenta, unico erede, proseguiva la tradizion familiare. Egli era, in verità, l'ideal tipo del giovine signore italiano nel XIX secolo, il legittimo campione d'una *stirpe* di gentiluomini

shaping it (and no discussion of D'Annunzio's occupation of Fiume). Burgwyn notes that D'Annunzio's first enunciation of "vittoria mutilata" appears in "La preghiera di Sernaglia" (October 1918), in which he declares: "Our victory, you will not be mutilated" ["Vittoria nostra, non sarai mutilata"]. I will return to this address below. Published in *Il libro ascetico della giovane Italia* (D'Annunzio, *Prose di ricerca*).

7 See, for example, Barilli 46. In response to allegations by Emilio Treves that he had desecrated the memory of the Italian defeat at Dogali by having his protagonist dismiss the dead soldiers as, "four hundred brutes who died brutally" ["quattrocento bruti, morti brutalmente"], in 1889 D'Annunzio wrote: "That sentence was uttered by Andrea Sperelli, and not by Gabriele D'Annunzio, and it sits well in the mouth of such a *monster*" (*Piacere,* 287, n. 2; 1229). Barbara Spackman suggests that, instead, Sperelli represents, "the culmination of a genealogical line rather than its rupture" (*Decadent Genealogies* 38).

8 For a critique of his status as precursor, see Spackman's chapter, "D'Annunzio and the Antidemocratic Fantasy" in *Fascist Virilities* 77–113.

> e di artisti eleganti, l'ultimo discendente d'una *razza* intellettuale. (emphasis mine; *Prose di romanzi* I 35–36)
>
> [Count Andrea Sperelli-Fieschi d'Ugenta, sole heir to the family, carried on its traditions. He was, in truth, the ideal type of young Italian nobleman of the nineteenth century, a true representative of a race [*stirpe*] of chivalrous gentlemen and graceful artists, the last scion of an intellectual line [*razza*].] (*The Child of Pleasure* 23)

While *stirpe* is aligned with legitimacy within a celebrated bloodline, *razza* surfaces to mark its final dissolution. *Razza* appears in *Il Piacere* as a signal for genealogical truncation. Its enunciation within the text brings about a concomitant structural truncation, as it brings to a close extended meditations on genealogy or progeny: *Il Piacere* lingers considerably less in the realms of heredity and reproductivity than its successors *L'Innocente* and *Trionfo della morte*. Sperelli's trajectory within the novel is propelled by overlapping quests for sexual, artistic, and spiritual stimulation which, ultimately, lead him back where readers first encountered him—alone (the only [re]birth celebrated being his own, however dubious) at the Palazzo Zuccari.[9]

Within the narrative economy of *Il Piacere*, the ultimate fruitlessness of Sperelli's regenerative meanderings is conveyed through the relationship of blood to the dyad of *stirpe–razza*. Rather than flowing along a genealogical course, in *Il Piacere*, blood, confined to the veins of the individual body that it inhabits, surges most often to signify not heredity but sexual arousal. It is a rhetorical container for the text's erotic discourse: for Sperelli, it is a "torrent" (78) and a "tumult" (82) that is reignited with sexual appetite (102).[10]

Andrea's non-reproductive unions with both Elena Muti and Donna Maria Ferres are characterized by the metaphorical exchange of vital fluids. At the height of Andrea and Elena's impassioned encounter on the latter's

9 For Nicoletta Pireddu, Sperelli's failure does not preclude the novel's formulation of a symbolic economy of art and the gift that defies mercantile logic ("Gabriele D'Annunzio" 175). Giorgio Bàrberi Squarotti reads *Il Piacere* as an extended meditation on the erosion of the aristocracy through the diffusion of bourgeois capital and Sperelli as a failed hero insofar as the decline of his romantic life parallels the degradation of aristocratic standards of art and beauty (77).

10 Barbara Spackman has identified this erotic discourse as the rhetoric of sickness, and has argued that "there is a movement in the trilogy from an eroticization of female sickness in *Il Piacere* to its total deeroticization in *Trionfo della morte*" (*Decadent Genealogies* 154).

sickbed, as the two repeat their seductive and subsequently plagiarized (with Donna Maria) refrain: "How I love you!" [...] "Again!", Elena's words are described as "so liquid and sensual." The narration proceeds: "Then both fell silent. He felt her presence flow and mix in his blood, until it became her life and her blood his life" (*Prose di romanzi* I 85).[11] When Andrea struggles later to convince Donna Maria of his short-term memory loss and accompanying refusal of Elena, to the query, "Who was that Elena?", Andrea responds:

> Non so; non ricordo. Non ricordo più nulla. Vi amo. Amo voi sola. Penso per voi sola. Vivo per voi sola. Non so più nulla; non ricordo più nulla; non desidero più nulla, oltre il vostro amore. Nessun filo più mi lega alla vita d'un tempo. Sono ora fuor del mondo, interamente perduto nel vostro essere. Io sono nel vostro sangue e nella vostra anima; io *mi sento* in ogni palpito delle vostre arterie; io non vi tocco eppure mi mescolo con voi come se vi tenessi di continuo tra le mie braccia, su la mia bocca, sul mio cuore. (313)

> [I do not know, I have forgotten. I remember nothing but that I love you. I love none but you. I think only of you. I live for you alone. I know nothing, I wish for nothing but your love. Every fetter that binds me to my former life is broken. Now I am far from the world, utterly lost in you. I live in your [blood] and in your soul; I *feel myself* in every throb of your pulse; I do not touch you, and yet I am close to you as if I held you in my arms, pressed to my lips, to my heart.] (270, emphasis D'Annunzio's)

The series of negations that introduce the moment of metaphorical transfusion perform on Andrea a sort of draining. Emptied of memory, reason, and desire, having severed the *filo* that bound him to Elena, Andrea enters Donna Maria through vein rather than vulva.[12] These metaphorical transfusions are embedded not within a procreative discourse, but within an erotic one whose height is not the evacuation of another fluid in the service of fertilization (as in *L'Innocente*) but, instead, a perpetuation of the site of arousal: blood exchanged is pleasure sustained.[13] Yet, the final link in the ideological chain

11 The entire paragraph is omitted from the English translation from which I cite. This translation is my own.

12 For an account of Andrea's possession of Maria, see Spackman, *Decadent Genealogies* 96–102.

13 *Il Piacere* addresses blood through an erotic economy of exchange, one that lies

of signification—wherein the blood of sexuality joins that of race in the act of procreation—occurs not in *Il Piacere*, but elsewhere.

Degeneration as *Svenamento* in *Trionfo della morte*

The tethering of circulation to stimulation in *Il Piacere* has shifted by *Trionfo della morte*, as its protagonist Giorgio Aurispa is more preoccupied by the role of blood in communicating a disquieting genetic code than by its arousal in love. If Andrea Sperelli is figured, at least initially, as the "champion" of an illustrious line, readers meet Giorgio as the final blow to his Casa Aurispa is being struck. Significantly, siege to the house is laid not by representatives of a new democratic state (as is the case with the oft-cited "grigio diluvio democratico odierno" of *Il Piacere*), but from within. Giorgio's anguished mother explains how his father has brought about the decline of the family:

> Aveva rovinate le terre, tagliato gli alberi, venduto il bestiame, così, alla cieca, alla prima occasione, al primo offerente. Ora incominciava a spogliare la casa, dove i suoi figli erano nati. Da molto tempo aveva messi gli occhi su quell'argenteria: argenteria di famiglia, antica, ereditaria, custodita sempre come una reliquia della grandezza di Casa Aurispa, conservata fino a quel giorno intera. (731)
>
> [He had ruined the property, cut down the trees, sold the stock blindly to any one—at any price. And now he had begun to strip the house in which his children were born. For a long time he had cast longing eyes on the silver, the family heirlooms, always jealously guarded as a relic of the ancient grandeur of the Aurispas, and preserved intact till this day.] (D'Annunzio, *The Triumph of Death* 79)

This economic threat is paralleled in the novel by a genetic one; indeed, within the *fin-de-siècle* Italian novel, the erosion of the aristocracy is often figured as both financial and biological.[14] In *Trionfo della morte, razza*

outside procreation and/or reproduction. Such a reading of *Il Piacere*'s erotic discourse fits nicely alongside Pireddu's essay on D'Annunzio's search for an aesthetics of squandering that defies a productive, utilitarian logic ("Gabriele D'Annunzio").

14 Bàrberi Squarotti hints at this connection in his reading of *Trionfo della morte*. Though his account focuses on Aurispa's intellectual and spiritual decline, he draws

bubbles to the surface yet again, this time ideologically fused to blood, as both contain the flawed heredity that ultimately provokes the protagonist Aurispa's suicidal and homicidal plunge.

"Death triumphs in this novel not because the flight from the flesh is total," writes Barbara Spackman. She continues, "Aurispa flees not only from the 'flesh' of the sexual act but from his own flesh and blood" (*Decadent Genealogies* 187). While Giorgio's mother is exempt from the "germi ereditati dal padre," Giorgio's paternal line is marked by genetic defect. Significantly, both *razza* and *sangue* characterize the relationship of these afflicted bodies to Giorgio, yet Giorgio's mother retains, "in her whole appearance an unmistakable air of breeding [*stirpe*]" (60). Similar to its formulation in *Il Piacere*, here *stirpe* contains the sunny potential of salubrious heredity, whereas *razza* teams up with *sangue* to signify genetic descent. Giorgio is horrified by Zia Gioconda: "This poor creature, he thought, who has sunk to the last depth of human degradation—this miserable, bigoted old sweet-tooth is connected with me by the insuperable tie of blood [*vincoli di sangue*]—she and I belong to the same race [*razza*]" (57). Similarly, Giorgio's brother Diego is distanced from Cristina (who instead "inherited much of the maternal grace") at the level of blood: "Would anyone believe they belonged to the same [bloodline] [*sangue*]?" (67),[15] and from Giorgio at the level of race: "[Giorgio], the favorite of fortune [...] who lived [...] as far removed from his family as a being of another race [*razza*]" (81).

At the behest of his mother, Giorgio seeks out his father in the flesh in order to settle precarious family finances, and the elder Aurispa laments:

> Invece, da qualche tempo, tutto mi va a rovescio, tutto; le disgrazie piombano una dopo l'altra; ho avuto perdite gravissime: tre cattive annate, di seguito: la malattia delle viti, il bestiame decimato; le rendite sono diminuite di più della metà; le imposte sono cresciute, enormemente... Vedi, vedi: queste sono le carte che ti volevo mostrare...[...] Bisognava mettersi in regola subito per evitare danni incalcolabili. (750)
>
> [Why [...], everything has gone from bad to worse lately. I have had heavy

a connection between economy and physiology. Instead of a *superuomo*, Aurispa represents, "the impossibility of saving himself from the economic and physiological degradation in which the world is involved with recourse to exclusivist and economic ideologies of bourgeois reality" (90).

15 Translation altered as indicated.

> losses—three bad years in succession—disease in the vines—the cattle dying—the rents reduced by one-half, the taxes enormously increased. See here, here are the papers I wanted to show you [...] They ought to be attended to at once if incalculable [damage] is to be averted.] (95)

Here, Giorgio's father holds up a mirror image of the scenario alleged by Giorgio's mother: instead of the former bearing responsibility for the sack of the Casa Aurispa, a ruthless mother nature and institutional foreclosure are to blame. Giorgio, whose pity momentarily quells his skepticism, immediately recalls a version of a primal scene:

> Un'immagine gli sorse dalla memoria, stranamente precisa, in un ricordo della fanciullezza:—il padre seduto accanto a una finestra, serio in viso, con la camicia rimboccata su un braccio ch'egli teneva immerso in un catino pieno d'acqua; e l'acqua che s'arrossava pel sangue fluente dalla vena aperta; e accanto, in piedi, il flebotomo che sorvegliava il flusso tenendo pronto l'occorrente per la legatura.—Un'immagine si associava all'altra; ed egli rivedeva le lancette luccicanti nell'astuccio di pelle verde, rivedeva la donna che portava via dalla stanza il catino pieno di sangue, rivedeva la mano legata da un nastro nero che s'incrociava sul dorso pieno e molle affondandovisi un poco. (750–751)

> [A scene out of his childhood rose with startling distinctness before him—his father seated near a window with a very grave face, his shirt sleeve pushed far up his arm, which he held over a basin of water; the water stained with blood which flowed from an open vein, and beside him the surgeon watching the flow of blood and holding the bandage ready for application. One picture called up another; he saw the gleam of lancets in a green leather case, a maid carrying away the basin full of blood; he saw the arm in a black silk sling tied across the broad back.] (96)

The economic ruin of the aristocratic line is figured as bloodletting, thus affirming the inextricability of *sangue* and *razza* within the novel. The soiled contents of his genetic line both exposed and emptied, Giorgio turns, like Andrea Sperelli before him, to face the sea. Yet Aurispa's displacement conditions not an isolated and invigorating convalescence, but a reunion with both his lover Ippolita, and an "experiment" in "com[ing] in contact with the inferior strata of his race [*razza*]" (215).

Like the "pinzòchera" Zia Gioconda, the churning swarms of devotees at Casalbordino are disfigured bodies, palpitating with disease and injury. These descriptions of popular religious ritual are some of the most insistent in the novel, their overwhelming density tempered only somewhat by the intermittent, ventriloquized voices that supplicate, "Viva Maria!" It is as if D'Annunzio, overflowing with imagery acquired both from Francesco Paolo Michetti's 1883 painting Il Voto and his first-hand observation of a similar scene in 1887, both of which images in turn embellished by his study of medical texts for *L'Innocente,* could not resist spilling the grisly surplus across the pages of the trilogy's final installment.[16] If Giorgio's father is bled by banks and blight, these peasants—amputees with fresh wounds, desiccated women, unable to conceive or lactate—represent the depths of such genetic depreciation.[17] Like Giorgio's father, these bloodied throngs, figured as ghouls in a house of horror, beg for charity:

> Un accattone comparve d'improvviso come se fosse balzato di sotterra; e tese la mano.
> —La carità, per amore della Madonna!
> Era un giovine, col capo fasciato da un fazzoletto rosso che per un lembo gli copriva un occhio. Sollevò il lembo e mostrò l'occhio enorme, gonfio come una borsa, purulento, in cui il battito della palpebra superiore metteva un tremolio orribile a vedersi.
> —La carità, per amore della Madonna!
> Giorgio gli fece l'elemosina; ed egli ricoprì la bruttura. Ma, poco oltre, un uomo gigantesco, sanguigno, monco d'un braccio, si trasse a metà la camicia per mostrare la cicatrice increspata e rossastra dell'amputazione. (898)

> [A beggar appeared suddenly before them as if he had sprung out of the earth. He held out his hand.
> 'Charity, for the love of the Madonna!'

16 See notes: Mondadori edition (n. 871: 1, 1330 and n. 885:1, 1332). For his consultation of medical texts, see Spackman, *Decadent Genealogies* 140, n. 23.

17 Spackman reads these scenes, "in which all imaginable illnesses and deformations of humankind are presented not as a projection of the father but as Ippolita's dream, as the unleashing of Pandora's box" (189). I approach the sanctuary scenes by way of an alternative route, which incorporates both the lengthy discussion of his father's disgusting organs and flesh cited by Spackman (188), and the relationship between economic and genetic decline represented throughout the novel through the interplay of *razza, sangue,* and *stirpe,* forged in particular by Giorgio's recall of his father giving blood in the wake of the latter's plea for money (cited above).

> He was quite young, and had his head tied up in a red handkerchief, a corner of which covered one of his eyes. He raised the flap and disclosed an eye swollen like a pouch and discharging, the eyelid having a nervous tremor perfectly horrible to witness.
> 'Charity, for the love of the Madonna!'
> Giorgio hastily gave him something, upon which he covered up the hideous sore. A little further on, they were confronted by a gigantic one-armed man, who drew off part of his shirt in order to exhibit the seamed and raw-looking scar of the amputation.] (220–221)

True to gothic form, just as Giorgio and Ippolita think they are safe, they are once again assailed:

> E tra la polvere apparve in confuso un viluppo di mostri. Uno dalle mani mozze agitava i moncherini sanguigni come se la troncatura fosse ancor fresca o mal catrizzata. Un altro aveva le palme munite d'un disco di cuoio e su quelle trascinava a fatica la massa del corpo inerte. Un altro aveva un gran gozzo grinzoso e violaceo che gli ondeggiava come una giogaia. Un altro, per una crescenza del labbro, pareva tenesse fra i denti un brano di fegato crudo. Un altro mostrava il volto devastato da una erosione profonda che gli scopriva le fosse nasali e la mascella di sopra. Altri mostravano altri orrori, a gara, con gesti violenti, con attitudini quasi di minaccia, come per far prevalere un diritto.
> —Ferma! Ferma!
> —Date l'elemosina!
> —Guardate! Guardate! Guardate!
> —A me! A me! (901)

> [Through the veil of dust there appeared a scrambling pack of monsters. One of them waved the stumps of his amputated hands, which looked raw and gory, as if the mutilation were quite recent; another had leather pads fixed to the pads of his hand, by means of which he painfully dragged along the lifeless mass of his body; a third had a monstrous goiter, wrinkled and purple, that flapped like a dewlap; another, by reason of an excrescence on his lip, seemed to be holding a piece of raw liver between his teeth; another one had his face so eaten away by a cancer that the nasal bone and upper jaw were laid bare; others displayed a variety of other horrors with infinite gusto, with violent, not to say menacing gesticulations, as if asserting a right.

'Stop! Stop!'
'Give us alms!'
'Look—look at me!'] (223)

The association of these peasants with the cursed Casa Aurispa is shored up through their figuring as representatives, blood-soaked, mutilated, and destitute, of these "human monstrosities, these remnants of a worn-out race [*una razza disfatta*]" (902).[18] Lucia Re emphasizes the "intense ethnographic" and "clinical tone" of the novel, pointing out that the "savage, atavistic, and animalistic behavior of the Abruzzese fanatical and monstrously deformed multitudes [*sic*] recall the Lombrosian anthropological style of the likes of Alfredo Niceforo's *L'Italia barbara contemporanea*" ("Italians and the Invention of Race" 13). In order to arrive at this feared reawakening of Ippolita's "male sacro," however, Giorgio must resolve his own relationship to the cadaveric bodies at the sanctuary.

Giorgio's confrontation with the devoted masses forces a momentary reconsideration of his refutation of Ippolita in the name of ascetic revival: "The sentiment of his love seemed to him refreshed and rejuvenated after all the strange and hideous experiences he had just gone through" (214–215). The imbrication of father and crowd through their respective physiological descriptions is supplemented by a figure for the exposure of the genetic code: the metaphorical bond between blood and racialized genes is brought to the surface, spewing forth from the bodies of both. The disquieting vision of the crowd thus provokes a reckoning with his father: Giorgio struggles to situate himself within these overlapping genealogies. "He had come in contact with the inferior strata of his race [*razza*], and the result had been nothing but unmitigated horror" (215). Yet the "horror" that he confronts is not simply that of the gruesome dismemberment of peasants, kept at a comfortable distance by the grace of inherited social status. The racialized continuum has been marked, and his *casa* faces a precipice. For Giorgio, what results from witnessing the crowd is a crisis in racialized subjectivity, which he resolves, like the fetishist, through a decision to "respect the veils." As Spackman writes about the trilogy: "[A]n unveiling that concludes veils are necessary is nothing other than the fetishist's disavowal. In differing forms, the novels of the trilogy narrativize the logic of fetishism, and the 'law' transgressed is the

18 These scenes of desperation launch Giorgio into an elaboration of his anxiety about Ippolita's epileptic relapse, which Spackman argues affirms Giorgio's ultimately unfulfilled desire for sexual and poetic priority (*Decadent Genealogies* 184).

law that constitutes sexual difference" (*Decadent Genealogies* 155). Moreover, given Bhabha's identification of a structural analogy between the fetishist's disavowal of sexual and racial difference, such an analogy is also at work in how *Trionfo*'s Giorgio approaches race.

"[His being did not have roots in that soil]; he could have nothing in common with this multitude" the narration proceeds, drawing a distinction between the physiological reality posited by the text ("his being did not have roots in that soil") and Giorgio's subjective experience of it ("he could have nothing in common with this multitude").[19] He did not, he could not: his existential search for origins leads him to confront his genealogy only to renounce it. The genetic blood that links Giorgio to the crowd here flows into another kind of liquid:

> Ora egli s'accorgeva che, volendo ritrovare tutto sé e riconoscere la sua vera essenza nel contatto immediato con la razza da cui era uscito, errava come chi volesse ricercar le cause della forma, della dimensione, della direzione, della velocità, della forza di un'onda marina nella massa acquea sottostante. Lo scopo dell'esperimento era fallito. Egli era estraneo a quella moltitudine come a una tribù di oceanidi; egli era anche estraneo al suo paese, alla terra natale, alla patria, com'era estraneo alla sua famiglia, alla sua casa. Egli doveva rinunziare per sempre a quella vana ricerca del punto fisso, dell'appoggio stabile, del sostegno sicuro. (892)
>
> [(Now he realized that, wanting to recover his entire self and recognize his true being in immediate contact with the race from which he was born, he erred like someone who wanted to find the causes for the form,

19 The "multitude" here refers to those floundering in Giorgio's declining gene pool. Following a similar course as the ideologeme of blood, at Fiume, the multitude will take on a positive character as a synonym for the heroic and unified people. Hardt and Negri have theorized the multitude—a critical concept in their call for an affirmative biopolitics (which they name "biopolitical production," in opposition to the oppressive "biopower"). "An internally different, multiple social subject whose constitution and action is based not on identity or unity (or, much less, indifference), but on what it has in common," the multitude challenges the conventional basis of sovereignty (which lies in the people, characterized instead by a setting aside of difference in favor of undifferentiated unity) (*Multitude. War and Democracy in the Age of Empire* 93–100). At Fiume, perhaps not surprisingly, D'Annunzio's multitude does not stand in radical opposition to sovereignty, but for an undifferentiated people who willingly submits itself to a sovereign (presumably to D'Annunzio himself).

dimension, direction, speed and force of a wave in the aqueous mass below it.) His experiment had failed utterly. He was as much a stranger to these people as if they were a tribe of South Sea islanders, as much an alien to his country and his native soil as he was to his family and his childhood's home. He must renounce forever the vain search for a fixed point, a stable and assured support.] (215)[20]

Race, here figured as *razza,* thus envelops both territory (*paese, terra natale, patria*) and biology (*famiglia, casa*). Refusing to anchor himself in race, which would require the embrace of a mortal genetic destiny, Giorgio affirms the futility of his subjective quest. Yet this refusal requires an oscillation:

> Ma perché dunque, volendo conservare la vita, non diverrebbe egli, a forza di metodo, così valido e così agile da abituarsi a rimanere in equilibrio pur tra quelle diverse impulsioni e a danzare pur su l'orlo del precipizio liberamente e arditamente? (892)
>
> [But since he so greatly desired to preserve his life, would he not, by force of practice, become sufficiently expert to maintain his balance amid the varying impulses, and tread freely and fearlessly on the edge of the precipice?] (215)

Here, the "precipice" is staged as one that separates the fatalist genetic code from the possibility of life. This image of suspension in hand, Giorgio resolves to maintain it in precisely the way the fetishist would:

> Ora intendi lo spirito acquistare il disgusto della verità e della certezza, se vuoi vivere. Rinunzia all'acuta esperienza. Rispetta i veli. Credi nella linea visibile e nella parola preferita. Non cercare oltre il mondo delle apparenze creato dai tuoi sensi meravigliosi. Adora l'illusione. (893)
>
> [You must train your mind to avoid truth and certitude if you would live—renounce all keen experience. [Respect the veils]. Look not beyond the world of appearances created by your own vivid imagination. Adore the illusion.] (216)

20 The parenthetical section is omitted from the English translation I have been citing; the translation is therefore my own.

Giorgio concludes that he must triumph over seeking to root himself genealogically (posited here as "truth" and "certainty"), and embrace the "illusion" of its inconsequentiality.

Yet, as readers of even the title know, ultimately, death triumphs.[21] The *precipizio* from the passage above returns at the novel's conclusion, and Giorgio, no longer content to teeter at the summit, plummets to his death, taking his lover with him. Giorgio and Ippolita ultimately "[crash] down headlong into death, locked in that fierce embrace" (315). Before the fatal leap, Giorgio imagines Ippolita's murder several times, and even comes close to submerging her in a watery grave (*Prose di romanzi* I 952–956). Perhaps the most significant of these moments for the purposes of this discussion comes toward the end of the novel, just as the couple nears the height of amorous inebriation. Giorgio beholds Ippolita, eroticizing the contrast between her browned flesh and her transparent wrists and effecting a kind of racialization through the aestheticization of skin:

> Sebbene la sua carne fosse bruna, d'un color d'oro caldo e opaco, ella aveva nei polsi una pelle estremamente fine, assai più chiara, d'un pallor singolare. Il sole aveva oscurata la parte delle braccia esposta; ma, di sotto, i polsi erano rimasti pallidi. E su quella finezza, a traverso quel pallore, trasparivano le vene esili ma visibilissime, d'un azzurro intenso, d'un azzurro pendente un poco nel violetto. [...] Ippolita disse, tenendo l'uno e l'altro polso:
> —Bacia!
> Egli ne afferrò uno, e col coltello fece l'atto di secarlo.
> —Taglia pure—sfidò ella.—Io non mi muovo.
> Nell'atto egli guardava fiso la delicata trama cerulea su quella pelle così chiara che pareva appartenere ad un altro corpo, a un corpo di donna bionda. E la singolarità l'attraeva e lo tentava esteticamente, suggerendogli un'immagine tragica di bellezza.

21 Most scholars agree that death in the novel is linked to D'Annunzio's Nietzchean and Wagnerian intertexts. Laura Wittman, writing about Aurispa's longing for the psychic cohesion afforded by death, discusses two important precursors: Nietzsche's Zarathustra, "the emblem of ascendant philosophy, in which death is seen as part of the cycle of eternal return: its truth is always present, and depends only on the individual will and its capacity not just to enter it but to celebrate it, thereby appropriating its power," and Wagner's *Tristan und Isolde*, according to which death is instead "the emblem of the self's dissolution" ("Mystical Insight and Psychology" 43–44).

—Questo è il tuo punto vulnerabile—disse egli sorridendo.—Il segno è palese. Tu morrai svenata. Dammi l'altra mano!] (1004–1005)

[Although her skin was brown in tone—a warm, pale gold—that on her hands was remarkably fine and very much whiter—peculiarly pale. The sun had burned her arms where they were at all exposed, but her hands remained white, and through the transparent skin the veins shone, fine but perfectly visible, of an intense blue, almost approaching violet. [...] Ippolita held out her two hands to him. 'Kiss!' she said.
He caught one of them, and made a pass with his knife as if to cut it off at the wrist.
'Cut away,' she said fearlessly. 'I shall not move.'
As he made the pass he gazed earnestly at the delicate blue lines under the skin, which was so white it seemed as if it must belong to another body, to a fair woman. This particularly attracted him, offered him an aesthetic temptation, suggesting to him an image of tragic beauty.
'That is your vulnerable spot,' he said with a smile. 'It is a sure sign—you will die of loss of blood. Give me the other one.'](303)

Ippolita participates willingly in this fantasy of bloodletting, and declares Giorgio's image of her death by *svenamento* (the cutting of veins) "bellissima":

(A)ppariva la moritura protendendo le braccia ignude alle cui estremità, dalle vene recise dei polsi, zampillavano e palpitavano due rosse fontane. E tra le due rosse fontane la faccia lentamente assumeva un soprannaturale pallore e le profondità degli occhi s'empivano d'un mistero infinito e su la chiusa bocca si disegnava la larva di una parola indicibile. D'un tratto i due getti cessavano. Il corpo esangue cadeva indietro, di schianto, nell'ombra. (1005)

[The picture rose up in complete detail before him. [A woman who was about to die appeared], her arms extended, and from her wrists, where the pulse arteries had been severed, two crimson jets sprang up and throbbed. Between these two crimson fountains the face slowly assumed a sepulchral pallor, the hollow eyes filled with infinite mystery, the wraith of an ineffable word hovered over the set lips. Suddenly the fountains ceased to flow, the bloodless body fell back into the shadow.] (303–304)

Giorgio's fantasy literalizes the metaphor: within Ippolita's veins runs a deep blue blood, which is only visible due to the extreme pallor of her un-suntanned flesh.[22] Giorgio describes her skin as, "so white it seemed as if it must belong to another body, to a fair woman." It is as if, even in the context of fantasy, Giorgio knows that hers is not the pure, "blue blood" of the aristocracy, but "plebeian blood" ("sangue plebeo").

It is of further note that the "two crimson jets" restore Ippolita to D'Annunzio's canon of female beauty that is marked almost obsessively by extreme pallor.[23] For example, Giuliana Hermil, wife of Tullio Hermil, protagonist of *L'Innocente*, languishes in her sickbed as Tullio beholds her:

> Quanto mi piacque! Come l'adorai, in quel momento! Come sentii che nulla al mondo vale la semplice commozione della bontà!
> Una bontà infinita emanava da quella creatura e mi penetrava tutto l'essere, mi colmava il cuore. Ella stava nel letto supina, rialzata da due o tre guanciali; e la sua faccia dall'abbondanza dei capelli castagni un poco rilasciati acquistava una finezza estrema, una specie d'immaterialità apparente. Aveva una camicia chiusa intorno al collo, chiusa intorno ai polsi; e le sue mani posavano sul lenzuolo, prone, così pallide che soltanto le vene azzurre le distinguevano dal lino. (375)

> [How beautiful she looked! How I adored her at that moment! I felt that nothing in the world could compare with her sweet and simple emotion. A sense of infinite goodness seemed to emanate from her and penetrate my whole being, filling my heart to o'erflowing. She was sitting in bed propped up by two or three pillows, and her face, framed in the

22 Montagu explains, "The term 'blue blood,' which refers to a presumed special kind of blood supposed to flow in the veins of ancient and aristocratic families, actually represents a translation from the Spanish *sangre azul*, the 'blue blood' attributed to some of the oldest and proudest families of Castile, who claimed never to have been contaminated by 'foreign blood.' Many of these families were of fair complexion, hence in members of these families the veins would, in comparison with those of the members of the predominantly dark-complexioned population, appear strikingly blue" (*Man's Most Dangerous Myth* 362).

23 For D'Annunzio, the pallor of female characters is figured as desirable both in and out of the sickbed. *Il Piacere*'s Donna Maria Ferres' sacralized beauty is figured through her white skin: "Maria had chosen that supernaturally white night to sacrifice [*immolare*] her own whiteness to his desire. All of the white things around, aware of the great sacrifice, waited to say *ave* and *amen* as the sister passed by" (303). Similarly, Elena Muti's shoulders are "as pallid as polished ivory" (42).

> loosened masses of her chestnut hair, had a look of extreme, almost ethereal delicacy; her hands lay listlessly before her, so white that the blue veins alone distinguished them from the sheet.] (D'Annunzio, *L'Innocente (The Victim)* 16–17)

In this way, Spackman's argument about female sickness as a constitutive part of D'Annunzio's erotic discourse is bound to a kind of racialization: the production of whiteness.[24] In *Trionfo della morte,* Ippolita's pallor belongs to "another body," as it is in her that the components of the constellation I have been tracing—*stirpe, razza,* and *sangue*—converge. And it is this convergence that marks her fate, first within the context of the homicidal daydream discussed above, and finally with the conclusion of the novel.

How did this fanciful draining make its way into the erotic poetics of the novel? It is certainly ideologically linked to a shift in eroticization of female sickness within the trilogy. In addition, as in the case of Giorgio's father and the Abruzzi crowd, the novel's logic of racialization binds sanguinity to economics. At an earlier point, as Giorgio beholds Ippolita, rather than her pallid splendor, he describes her genetic flaws:

> Non erano belli i piedi nudi ch'ella a volta a volta scaldava su la ghiaia e rinfrescava nell'acqua; erano anzi difformati nelle dita, plebei, senza alcuna finezza; avevano l'impronta manifesta della bassa stirpe. Egli li guardava intentamente; non guardava se non quelli, con uno straordinario acume di percezione e di esame, come se le particolarità della forma dovessero rivelargli un segreto. E pensava: 'Quante cose impure fermentano nel suo sangue! Tutti gli istinti ereditarii della sua razza sono in lei, indistruttibili, pronti a svilupparsi e ad insorgere contro qualunque constrizione. Io non potrò mai far nulla per purificarla. Io non potrò se non sovrapporre alla realtà della sua persona le figure mutevoli dei miei sogni, ed ella non potrà se non offrire alla mia ebrezza solitaria i suoi indispensabili organi...' (915)

24 The whiteness of Italians was a subject of debate in the positivist anthropological works of Cesare Lombroso and Giuseppe Sergi. See, for instance, Lombroso, *L'uomo bianco e l'uomo di colore*; Sergi, *Le varietà umane*. For a history of the construction of whiteness in Italy, see: Giuliani and Lombardi-Diop, *Bianco e nero*. In her study of Italian advertising under fascism, Karen Pinkus examines discourse surrounding curative baths and sun-tanning (175–176).

[They were far from beautiful, those bare feet of hers, which she alternately warmed on the sand and cooled in the water; the toes were misshapen, plebeian, wholly without refinement, and bore the unmistakable traces of low origin [*stirpe*]. Giorgio observed them attentively, looked at nothing else, as if the details of their shape could afford him the clue to some mystery. 'What a host of impure things ferment in her blood,' he thought; 'all the inherited instincts of her race [*razza*] are handed down in her, indestructible, ready to develop themselves and rise up against any restraining influence. Never shall I succeed in purging her of that. All I can do is superimpose on her real character the changeful images of my fancy, and she can offer nothing to my lonely passion but the sensual instruments of love.] (235)

The novel's genetic nucleus is thus projected onto the figure of Ippolita, whose contaminated blood manifests itself in physical and psychological deformation, the stamp of her low extraction. No longer bearing the potential that it represented in Giorgio's mother, in Ippolita, *stirpe* appears as its inverse when aligned with *razza* and *sangue*. Giorgio's resignation ("Never shall I succeed in purging her of that") emerges both in relation to her epileptic possession and as a result of the shift in the rhetorical weight of blood: from an entrenchment in the poetics of non-reproductive eroticism in *Il Piacere*, in *Trionfo della morte* blood sexualized becomes the locus of the novel's genetic discourse (Spackman, *Decadent Genealogies* 190).

Ippolita's sterility may thereby be read as an ideological and narrative imperative, as it forces Giorgio to confront his own relationship to desire and reproduction:

Pensò: 'Ella è sterile. Il suo ventre è colpito di maledizione. Ogni germe vi perisce come in una fornace ardente. Ella inganna e delude in me, di continuo, il più profondo istinto della vita.' L'inutilità del suo amore gli apparve come una trasgressione mostruosa alla suprema legge. – Ma perché dunque il suo amore, non essendo se non una lussuria inquieta, aveva quel carattere di fatalità ineluttabile? Non era l'istinto di perpetuazione il motivo unico e vero d'ogni amor sessuale? (916)

[He thought: 'She is sterile. Her womb is cursed. Every seed perishes there as in a blazing furnace. She repeatedly betrays and disappoints my deepest vital instinct.' The uselessness of her love appeared to him as a monstrous transgression of the supreme law. Why then did his love, which was nothing but a restless lust, have that character of inevitable fatality? Wasn't the instinct of perpetuation the only true motive of every sexual love?][25]

The novel has already provided the answer to this question, through its configuration of the ideologemes of *stirpe, razza,* and *sangue.* Perpetuation is the charge only in the case of "the most superior beings," none of which populate the text. As Giorgio remarks, recalling Nietzsche's *Zarathustra* (Re, "Italians and the Invention of Race" 13), and reflecting upon his confrontation with the crowd:

> La specie umana aveva dunque un fondo interamente inerte che permaneva sotto le ondulazioni delle zone mobili superiori. Il tipo ideale dell'umanità non era dunque nel lontano futuro, non era al termine ignoto di un periodo progressivo; ma poteva solo manifestarsi alla sommità delle onde, negli esseri più elevati. (892)

> [T]he human species had an absolutely immovable basis unaffected by the undulatory motion of the upper zones. Therefore the ideal type of humanity was not to be looked for in the distant future, at the end of an unknown term of progressive evolutions; it could only manifest itself on the crest of the waters—among the most superior beings. (215)

The Darwinist imperative offered by the novel (and through Giorgio) requires not only the "instinct of perpetuation," but also its refusal in the face of economic and biological degeneration. Carlo Salinari sketches an identikit of D'Annunzio's version of Nietzsche's *Übermensch.* Within D'Annunzio's *Superuomo,* he claims, lie: "power [*potenza*], war, glory, scorn for plebes, an aristocratic conception of the world, the idea of Rome and the mission of Italy, and the cult of beauty" (64).[26] Salinari suggests that it is this convergence of characteristics that finds expression in Giorgio Aurispa as his desire for Ippolita culminates in violence, and, ultimately,

25 This section is omitted from the English translation I have been citing. The translation is my own.

26 For a detailed monograph on D'Annunzio's *superuomo,* see Vettori. Several readings of *Le vergini delle rocce* address the Nietzschean intertexts. See: Pireddu, "Gabriele D'Annunzio"; Spackman, *Fascist Virilities*; Re, "Gabriele D'Annunzio's novel *Le vergini delle rocce*; Schnapp, "Nietzsche's Italian style."

an aestheticized death (66).[27] Yet the fetishistic logic of the text's genetic discourse complicates this reading of *Trionfo* in general and its protagonist in particular. Aurispa's relationship to the racialized masses at Casalbordino are not characterized solely by "scorn" or an "aristocratic conception of the world." Instead, his position with relation to the crowd is linked to Ippolita, upon whom Giorgio's relationships to race, desire, and reproduction are projected. As Lucia Re explains, "Ippolita as a woman is [...] literally associated with the racial inferiority of the primitive, animalistic multitudes over which Giorgio must elevate himself." ("Italians and the Invention of Race" 13). In spite of Giorgio's avowed refutation of desire for a sterile woman with "plebeian blood," the novel's genetic discourse authorizes both this desire and Giorgio's decision to truncate two flawed genetic pools through suicide and homicide.

As the novel alleges, bad blood makes for bad genes. Whereas, in the passage cited above, in which Giorgio confronts his relationship to the crowd, likening himself to a "marine wave" and the crowd to the "aqueous mass below," here, the crest of the wave represents the unattainable. The liquid connection between the infected multitudes and Giorgio is thereby affirmed as Giorgio, rather than teetering at the top of the genetic wave and representing a "superior being" is instead engulfed by the (blood red) sea.[28]

27 For Pireddu, in D'Annunzio's later autobiographical works, and in particular in *Notturno* (1921), death is linked to D'Annunzio's aesthetic project, as it becomes a vehicle through which an inimitable life is achieved. In these writings, D'Annunzio will explore death as "a form of expenditure that cannot be appropriated or assimilated by anyone else. [...] [In staging] the death of the act of giving through the death of the giving agent, D'Annunzio aims at a creative process no longer lacerated by the distinction between gift and debt" ("Gabriele D'Annunzio"178–179).

28 Spackman draws a connection between crowd and sea in D'Annunzio's representation of the sea at Schifanoia in *Il Piacere*. She contends that *Il Piacere*'s sea is a version of Baudelaire's figure for the crowd as an aqueous movement (which allows for the multiplication of the protagonist's consciousness) in *Le peintre de la vie moderne*. For Spackman, this effect is produced in Sperelli through his contemplation of the sea: "D'Annunzio's sea, like Baudelaire's crowd, is a reservoir of electricity that acts upon and illuminates the subject who will record these jolts as poems or sketches." She proposes that the crowd expelled from *Il Piacere*'s scene of convalescence emerges in *Trionfo della morte* in precisely this scene at Casalbordino (*Decadent Genealogies* 53). If in *Il Piacere* the sea at Schifanoia stands in for the crowd, in an earlier text, "La vergine Orsola" (*Novelle della Pescara,* 1884–1886), D'Annunzio stays closer to Baudelaire's formulation, though he inscribes this watery movement within an erotic discourse, as it occasions Orsola's sensualized contact with various body parts: "In the church, the crowd was immense

From eroto-poetics to bio-poetics, a shift in the rhetorical deployment of blood has occurred from *Il Piacere* to *Trionfo della morte*. In *Il Piacere*, as it is confined to a non-reproductive erotic discourse, blood inaugurates the eclipse of a *razza*. In *Trionfo*, this *razza*, which consists of blood both exposed and flowing, is figured as already "undone." The site of this transfer is *L'Innocente*, the second of the trilogy and the only novel of the three to thematize the scene of racialized parturition. Crucially, this shift occurs as Giuliana Hermil gives birth to Filippo Arborio's son, the biological "intruder."

Rhetorical Transfusion in *L'Innocente*

The thematic and ideological focus of *L'Innocente*—deemed by Giorgio Bàrberi Squarotti the worst of D'Annunzio's novels (78)—is adulterous reproduction, staged as a racial intrusion. This narrative, written in the first person and staged as a confession, is the only one of the trilogy to offer an account of female parturition.[29] Protagonist Tullio Hermil's wife Giuliana gives birth to an adulterine baby boy, Raimondo, occasioning both her extended convalescence and Tullio's infanticide.[30]

Tullio Hermil is inserted, like Giorgio Aurispa after him, by way of *razza* into a decadent family line. Yet Hermil's primary preoccupation is not with his genetic make-up, however deficient.[31] Instead, he positions

under the forest of palms. Orsola was separated from Camilla by one of those currents that form irresistibly among the popular masses; she remained alone in that flow, among all that touching, among all of that shoving and breathing. She tried to open a breach: her hands encountered a man's back, and another pair of tepid hands whose touch troubled her" (D'Annunzio, *Tutte le novelle* 100).

29 The novel opens: "To go before a judge, and to say to him: 'I committed a crime. That poor creature would not have died if I had not killed it. I, Tullio Hermil, killed it" (360).

30 For Goudet, Hermil's homicidal leanings do not bar him from representing D'Annunzio's passage from the "aesthetic hero" (Sperelli) to the "ethical hero" (*D'Annunzio romanziere* 85).

31 Hermil remarks, much like Aurispa after him: "How many times was I, an ideologue, analyst, and sophist in a decadent era, pleased to be the descendant of that Raymond Hermil De Penedo who at Goletta performed prodigious acts of valor and ferociousness under the eyes of Charles V! The excessive development of my intelligence and my multi-faceted spirit could not have modified the basis of my substance, the hidden substrate in which all of the hereditary characteristics of my race [*razza*] were inscribed" (503).

himself with regard to flawed prospective heredity. His biological paternity affords him no male heirs, and Giuliana's male child stands to usurp Tullio: "And the boy, who was no son of mine, would grow up protected by her and by her assiduous care; would grow robust and handsome; would become self-willed as a little despot, lord it over my whole house" (194). It is within this context that the erotic content of blood from *Il Piacere* first encounters race, eventually enabling their fusion within the genetic discourse of *Trionfo della morte*. Significantly, parturition is the locus of this encounter.

Racialization marks Tullio's suspicion of Giuliana's infidelity. As such, her whiteness, which I mentioned above has to do with the trilogy's aestheticization of female sickness, is reconsidered by Tullio. Rather than representing her beauty, her pallor may mask her betrayal. In a paranoiac conversation with himself, Tullio remarks: "You know—none better—that Giuliana's skin is extraordinarily white. She is as *pallid as her gown*. The sacred epithet might well mask a profane interpretation" (emphasis D'Annunzio's; 47). Furthermore, Tullio's subsequent discovery of Giuliana's adulterous pregnancy occasions the first and primary deployment of *razza* within *L'Innocente*. Haunted by her lover Filippo Arborio, Tullio figures his response to his nemesis as physiological:

> Al paragone di Federico, la figura di quell'uomo, così fine, così nervosa, così feminea, si rimpiccioliva, s'immiseriva, diveniva spregevole per me ed ignobile. Sotto l'influsso del nuovo ideale di forza e di semplicità virile, ispiratomi dall'esempio fraterno, io non soltanto odiavo ma disprezzavo quell'essere complicato ed ambiguo che pure apparteneva alla mia stessa razza e aveva comuni con me alcune particolarità di costituzione cerebrale, come appariva dalla sua opera d'arte. (498)
>
> [Compared with Federico, the figure of the other man—so over refined, so nervous, so emasculate—dwindled to miserable proportions, became offensive and ignoble in my eyes. Under the influence of the new ideal of strength and manly [*virile*] simplicity inspired in me by my brother's example, I not only loathed, but I despised this complicated and ambiguous being notwithstanding that he belonged to my own strain [*razza*], and that we had certain intellectual peculiarities in common, as was evident in his literary work.] (160–161)

Tullio will counter the threat posed by Filippo and Giuliana's union with physical robustness modeled upon his virile brother, Federico. How, then, to read this deployment of *razza*? Purging the biological language and instead taking it for granted as "type," it posits a kind of literary genealogy, wherein Tullio is bound to Filippo through purely literary production. Yet Tullio's mention of a shared "cerebral constitution" complicates such a reading, ensuring that *razza* carries with it some physiological significance. Nevertheless, the emergence of such a formulation in the wake of Tullio's discovery of Giuliana's scandalous pregnancy suggests that what is at stake for Tullio is a physiological, if not racialized, intrusion. Indeed, his rejection of the child is staged at the level of blood:

> Non si trattava d'un rimpianto, d'un rimorso, d'un ricordo indistruttibile, d'una qualunque più amara cosa interiore, ma di un essere vivente. Il mio avvenire era legato a un essere vivente d'una vita tenace e malefica; era legato a un estraneo, a un intruso, a una creatura abominevole contro di cui non soltanto la mia anima ma la mia carne, tutto il mio sangue e tutte le mie fibre votavano un'avversione bruta, feroce, implacabile fino alla morte. (505)
>
> [There was no question here of regret, remorse, of an indelible memory, of any mental burden however heavy, but of a living being. My future was linked with a creature imbued with a tenacious and malignant life; was fettered to a stranger, an interloper, an abhorrent creature against who not only my spirit, but my flesh, every drop of blood, every fibre of my body, rose up in loathing—brutal, fierce, implacable till death.] (168–169)

The threat the child poses to Tullio's heredity is thus inextricable from the novel's poetics of blood. Consequently, following an immunitary logic, as Tullio imagines an abortion of the fetus, he stands to *rinascere* [be reborn] (*Prose di romanzi* I 532).

The metaphorical exchange of vital fluids in *Il Piacere* is literalized in *L'Innocente*, as Tullio envisions Giuliana's adultery: "possessed by another, having received an excretion from another, she carries in her womb another's seed" (478).[32] This liquid gives way to another sort during parturition, as Giuliana suffers a hemorrhage that nearly kills her and occasions her extended convalescence: "'Is it born? What if she is dead—if both are dead—mother and child?' [...] Then I suddenly had a vision of Giuliana [in the middle

32 Tullio's inner dialogue is omitted from the English translation I have been citing. The translation is my own.

of a red lake], gasping out her last breath—hemorrhage'" (240).[33] Shortly thereafter, Tullio contemplates her:

> Considerando il suo funereo pallore di cera, io vedevo ancora quelle macchie di sangue, tutto quel povero sangue sparso che aveva inzuppato i lenzuoli, attraversato i materassi, arrossato le mani del chirurgo. 'Chi le renderà tutto quel sangue?' (568)
>
> [As I considered her deathlike pallor, I could still see those bloodstains, all of that poor bloodshed that had drenched the sheets, stained the mattress and the hands of the surgeon. I thought: 'Who will restore all of that blood to her?'][34]

Given the way in which the text presents the fetus as a physiological intrusion, "poor blood," may be inscribed somewhere between a rhetoric of compassion and a logic of racialization that posits a line of continuity between one vital fluid (*seme*) and the next (*sangue*). The "poor" blood may thus belong to Giuliana, object of Tullio's renewed affections, or Arborio, source of both seed and (flawed) hereditary blood.

It is precisely this blood, spilled during labor and now contained within the body of the newborn infant, that Tullio seeks to supplant when he answers his query, "Who will restore all of that blood to her?" with a fantasy of transfusion, "How cheerfully would I transfer half of my blood to those veins!" (244). Giuliana's health will require, after the evacuation of Arborio's blood during labor, a rush of Tullio's vital fluid:

> Io non d'altro dovrei occuparmi che della guarigione di Giuliana. Non mi moverei di qui, sarei il più assiduo e il più dolce degli infermieri, riuscirei a compiere la trasfusione vitale, a compiere il miracolo, per forza d'amore. Ella non potrebbe non morire. Ella risorgerebbe a poco a poco, rigenerata, con sangue nuovo. (570)

33 For a discussion of Giuliana's convalescence, see Spackman, *Decadent Genealogies* 156–157.

34 The original reads: "'È nato? E se ella fosse morta? Se ambedue fossero morti? la madre e il figlio? L'emorragia, il sangue…' Vidi il lago rosso, e, in mezzo, Giuliana boccheggiante" (566). Here and above, I have modified the English translation, as D'Annunzio's insistence on the ubiquity of postpartum blood is omitted. The English translation condenses the scene, as Tullio remarks here (rather than two paragraphs later, as in the original): "How cheerfully would I transfer the half of my blood to those veins!" (244).

> [I should have nothing to think about but Giuliana's recovery. I would not move from her bedside, would be the gentlest and most assiduous of nurses, would accomplish the miracle, would bring her back to life [*trasfusione vitale*] by sheer force of love. It would be impossible for her not to get well; she would emerge out of the depths little by little, regenerated, with blood renewed.] (244–245)

Yet, if Tullio sees a possibility for Giuliana's regeneration through transfusion, no such possibility exists for the newborn. A product of Arborio's seed and bearer of his blood, baby Raimondo must thereby die. *L'Innocente* stages parturition as the scene where race and blood collide; the novel itself might be read as the site of a transfer (or transfusion) between the erotic rhetoric of blood at work within *Il Piacere* and the genetic one at work in *Trionfo della morte*. What happens to the convergence of blood and race staged in the *Romanzi della rosa* when the bodies in question become soldier-patriots and the naturalized terrain is contested? D'Annunzio will take up this relationship again on the shores of Fiume some twenty years later, this time figuring bloodshed as the necessary prerequisite for the regeneration of the race and the production of *vital subjects*.

The Sacralization of Blood in Fiume

Described variously as either blatantly imperialist or as a model for Third World liberation movements, D'Annunzio's occupation of the formerly Austro-Hungarian city of Fiume (today Rijeka, Croatia) from September of 1919 to January of 1921 has nevertheless unequivocally played a significant role in the historiography of Italian fascism, particularly with regard to how his spectacular politics fused the messianic with the popular in ways that are read to prefigure Mussolini.[35] In addition to his oratorical style, which engaged listeners in a politics of spectacle, among the D'Annunzian

35 For a discussion of this relationship that draws a direct line of connection between D'Annunzio's style at Fiume and that of Mussolini's fascism, see G. Mosse. See also Spackman, "*Il verbo (e)sangue*." For a history of D'Annunzio's Fiuman campaign that problematizes several ideologically rigid readings within the historiography of Fiume and for a sustained account of the relationship between D'Annunzio and Mussolini, see De Felice. Michael Ledeen, who follows Mosse's lead on the relationship between D'Annunzio's style and that of the Duce, remarks that on the eve of the occupation of Fiume, while Mussolini was merely "an aspiring politician," D'Annunzio was already, "an

inventions at Fiume most frequently cited as proto-fascist include the rally cry, "Eia, eia, eia, alalà!," the dressing of the Fiumian *arditi* (assault soldiers) in black shirts, and the frequent addresses to the populace from the balcony of a public square.[36] For eighteen months, D'Annunzio and a group of military defectors, adventurers young and old, journalists and artists occupied this city in northern Dalmatia with the goal of "redeeming" an ethnically Italian territory by annexing it to the Kingdom of Italy, whose claim to the city had been rejected at the Paris Peace Conference earlier that year. D'Annunzio's occupation soon achieved a sort of legendary status, both because of the libertine revelry that was rumored to accompany this ostensibly military operation and because a variety of colorful personalities, from wireless pioneer Guglielmo Marconi to futurist artist and writer F.T. Marinetti to Mussolini himself, were drawn to Fiume.[37] In addition, the ability of a poet to lead a lasting military occupation that threatened to destabilize the delicate balance of power that had recently been restored among Western powers was indeed an unprecedented event.[38] In terms of the trajectory of D'Annunzio's own aesthetic production, Fiume also constituted a pivotal moment.[39]

As this book is concerned with tracing the rhetorical imbrication of race and (re)productivity in Italy before their consolidation under fascism,

established national hero" (87). For D'Annunzio-as-Imperialist, see: Becker, *Nationalism and Culture*. For D'Annunzio-as-Third-World-Revolutionary, see Ledeen.

36 For a list of these and other innovations at Fiume that were to be appropriated by Mussolini, see Salaris 10.

37 For a detailed depiction of life in Fiume as recounted by artists and officials (among whom Mario Carli, Giovanni Comisso, Alceste DeAmbris, and Guido Keller), who took up the cause alongside D'Annunzio, see Salaris. For an account of Giuglielmo Marconi's historic visit, D'Annunzio's commemorative speech, and how D'Annunzio's speeches at Fiume as a whole construct listening, writing, and sacrificing subjects in ways that are indebted to the wireless technology developed by Marconi, see "D'Annunzio and the Marconigram: Crowd Control at Fiume" in Campbell, *Wireless Writing in the Age of Marconi*.

38 Ledeen examines in particular British and American responses to D'Annunzio's occupation. Burgwyn recounts an exchange between French Prime Minister Clemenceau, British Prime Minister Lloyd George, American President Wilson, and Italian Prime Minister Orlando upon the departure of the Italian Prime Minister from the Paris Peace talks, in which the three former expressed their affection for Orlando. Orlando allegedly replied, "You may still be fonder of me next week, when you may well be confronted with D'Annunzio in my place." Cited in Burgwyn 280.

39 Nicoletta Pireddu has argued that, at Fiume, D'Annunzio continued his meditations on the aesthetic economy of the gift by exploring the degree to which art could arouse popular consciousness and collective action ("Gabriele D'Annunzio" 175).

D'Annunzio's Fiume campaign is a critical site of analysis, as he simultaneously avails himself of anti-nationalist and anti-colonialist rhetoric. Indeed, it is within this semantic field that a distinction between "rhetoric" and "plain speech" risks reducing the complexity of D'Annunzio's discursive politics to one or another moral position.[40] Some readers may be tempted to dismiss D'Annunzio's call for a "League of Oppressed Nations" as "pure rhetoric," an oratorical smokescreen that obscures his "true" politics of racialized occupation. Others, eager to dispute readings of D'Annunzio's politics as proto-fascist and figure him instead as a revolutionist visionary, may be inclined to bracket the explicitly racialist and colonialist elements of his speeches as mere rhetorical flourish on an otherwise morally sound politics of anti-oppression. Either position, however, requires imposing static ideological boundaries on a text that skillfully dodges such attempts at classification. Instead, I draw from my discussion of D'Annunzio's earlier novels in order to address the complexity of race as a signifier within his Fiuman speeches. In so doing, I take seriously lexical inconsistencies, rather than bracketing them as ideological anomalies.

In D'Annunzio's *Romanzi della rosa* trilogy, race requires an ever-shifting relationship between *razza, stirpe,* and *sangue.* Though *Trionfo della morte* comes the closest of the three to thematizing a racialized collectivity, its reach is confined to the Abruzzi countryside.[41] In that novel, blood aligns itself with race (figured as both *stirpe* and *razza*) in order to signify genetic decline. As Lucia Re writes:

> The pervasive use of the idea of race in the novels [*Trionfo della morte* and *Le vergini delle rocce*] and elsewhere in D'Annunzio's work at this time, however, is still essentially divisive; it functions to reinforce and emphasize class and gender difference and conflict, rather than to transcend or obfuscate them. ("Italians and the Invention of Race" 13)

What happens to this constellation of terms when the stakes shift, both with regard to genre (from "high" literature to "popular" political address) and in terms of ideological project (from individual genius and genealogical degeneration to imperialist military occupation)?

40 For an eloquent and thorough discussion of the complex relationship of rhetoric and ideology to fascism, see Spackman, *Fascist Virilities* x.

41 The area formerly known as "Abruzzi" embraces the modern-day regions of Abruzzo and Molise, which were separated in 1963.

The bodies that D'Annunzio produces in his *Romanzi della rosa* trilogy—including those of his would-be protagonist-heroes—are hardly apt to brave the harsh winds and waters of Adriatic conquest. In his speeches and writings at Fiume, D'Annunzio refigures the relationship between physiological and economic decline in *Trionfo della morte*; for this later project, corporeal mutilation is mobilized not to signify the undoing of a race, but, at first glance paradoxically, in order to defend and extend its integrity. As George Mosse has remarked with regard to D'Annunzio's Fiuman campaign, "Regeneration was [...] linked to death and sacrifice" (37). Blood abounds in Fiume as a sacralized fluid that legitimates territorial expansion in the name of an Italian race.[42] Running parallel to the extension of the biological and territorial stakes at Fiume is a concomitant lexical inflation: new terms—*popolo, gente, patria*—inhabit these later texts. Within the sacralized rhetoric of Fiume, *popolo, gente, razza,* and *stirpe* are deployed in the service of the *patria*.[43] These four terms are used almost interchangeably; the result is that the relation of each to one another is subordinated to their collective relationship to the blood-soaked territory that is the basis for the *patria*.[44] It is as if D'Annunzio discovers "the population" at Fiume—in his

42 A social historian might explain this shift by highlighting D'Annunzio's participation in World War I, which, through the trials of trench warfare, quelled regional cultural-linguistic differentiation and cemented instead new forms of national allegiance. Barbara Spackman suggests that D'Annunzio's Christological rhetoric at Fiume emphasizes, rather than a prescriptive program of violence (Marinetti, Mussolini), "a violence suffered rather than perpetrated," and she reads the sacralized blood of Fiume, as "not the blood of a race, nor a blood after which to thirst, but the blood of already fallen martyrs" ("*Il verbo (e)sangue*" 222).

43 D'Annunzio's arrival at Fiume occurs after a literary detour through Africa. In 1905, he published *Più che l'amore*, a drama about protagonist Corrado Brando's tragic *mal d'Africa*. In 1911, on the event of the Italian invasion of Libya, D'Annunzio published his *Canzoni delle gesta d'Oltremare*. Both are rhetorical sites from which he will draw at Fiume, in particular with his formulation celebrating ancient Roman imperialism, "Teneo te, Africa." For a reading of these works alongside his speeches at Fiume, see: "D'Annunzio. Il superuomo, l'Africa, e l'eredità della stirpe latina," in Tomasello, *L'Africa tra mito e realtà* 69–82.

44 The convergence of these four terms positions them, as an ensemble, more readily with regard to blood and territory (and, thereby, to *patria*) than in relation to one another. There are, however, some significant moments when the terms diverge. *Popolo* and *stirpe* are mobilized to invoke the present-tense collectivity of an oppressed population, be it "white," "of color" (of the League of Oppressed Nations D'Annunzio proclaims, "It gathers together the white races and the races [*stirpi*] of color," 156; "It embraces the oppressed races [*stirpi*]," 313; and, "Meanwhile, in opposition to the League of Nations,

Romanzi della rosa trilogy, and elsewhere, D'Annunzio is concerned with the genetic fate of individual and collective bodies, yet when he reaches the Adriatic, these bodies have become *vital subjects*—a naturalized and nationalized collectivity. If *popolo, gente,* and *patria* are lexical novelties at Fiume, the signifier *moltitudine* is not. In the Abruzzi of *Trionfo della morte,* the multitude resembled an unruly and genetically corrosive crowd; at Fiume, it is recast in a positive light as synonymous with a productive and patriotic collectivity.

Disseminated as newspaper editorials, public orations, addresses to guests, officials, soldiers, and citizens from January of 1919 to January of 1921, D'Annunzio's Fiuman writings produce a similar relationship between biological and territorial belonging to those in both Franchetti's southern and colonial texts and in Mantegazza's address to the Monza electorate.[45] The blood of soldiers soaks the soil, marking and naturalizing the borders of the nation-state: "The barrier that divides what is ours from what does not belong to us is marked by the martyrs' blood" (D'Annunzio, *La penultima ventura* 276).[46] As Alberto Banti notes in his seminal study of the rhetoric

that conspiracy of privileged thieves and swindlers, we form the League of Fiume, which welcomes all of the representatives of all of those peoples who today suffer oppression and who see the living fibers of their national territories brutally mutilated," 231), or "Italian" (D'Annunzio hails his *Italiani* as, "a victorious people that wants and still knows how to win," 18; and he queries, "What does the true Italian people want, who are they with?" 54). *Razza* and *gente* are mobilized in the service of hereditary genius ("The people rediscovered the best imprints of the race [*razza*] in order to mold themselves accordingly" (64–65); "My young brothers, flowers of our chosen race [*razza*], at Fiume let us repeat a Roman saying: 'Here, we will remain excellently'" (209); for "the genius of the race" see 91; 296; "we are a Latin people [*gente*]," (153); "a genius of our people [*gente*]" (256)) and/or biological specificity ("a man who emerged from my old Abruzzi race [*razza*]" (47)). Yet each of these terms and the bodies to which they refer are bound to the sacralized blood shed within the pages of the Fiuman discourses. These and all subsequent translations of the Fiuman writings and speeches are my own. The page numbers quoted are taken from the original Italian in D'Annunzio, *La penultima ventura*.

45 These writings and addresses thereby anticipate and accompany D'Annunzio's occupation of Fiume, which ran from September 1919 through January 1921.

46 Examples of national territory bathed in the sacred blood of mutilated and dead soldiers abound: "Whomever dismisses you, whomever repudiates you, whomever betrays you knows that you are living creatures, peoples [*genti*] and cities, men and stone? If men bleed, stones restore his soul" (18); and: "Have you forgotten us, brothers? Our numbers are few, the survivors of a battered Italianness; few, but incredibly tenacious. Do you condemn us, brothers? Your condemnation is signaled by the ruins that litter our small, quadrangular Italic city. Your condemnation hangs above seven hundred Italians,

of blood bonds around the Risorgimento, "The constitutive elements of the national community [...] are blood, which binds a succession of generations to their land; a common tradition; and a common language" (*La nazione del Risorgimento* 62). These ties are still at work in D'Annunzio's Fiume, and D'Annunzio forges a further connection between labor productivity and biological reproductivity. *Italia* is figured as a prolific mother: "I repeat that this Mother of grain and of heroes, this woman warrior of the Solstice between sea and alp, still has among all other female generators the most fecund womb," who sheds blood as she gives birth (15–16). Her heroic sons are, "the lastborn of a bloody mother," who in turn shed their blood in her defense (74).[47] In a speech commemorating Italy's entrance into World War I, when "Rome became Roman once again," D'Annunzio recalls:

> L'Italia aveva partorito il suo futuro con uno spasimo atrocissimo; aveva ansimato prima di assalire; aveva sanguinato prima di combattere. Nelle ultime notti, le grida della moltitudine sembravano grida d'implorazione verso un dio redentore. Ed ecco taceva, quando la sua gente cominciava a morire sul suolo, quando la sua gente cominciava a morire sul mare, quando il suo sangue cominciava a scorrere, quando il suo miglior sangue cominciava a fumare davanti a una grandezza invisibile che era la sua grandezza promessa. (60)[48]

> [Italia had given birth to her future with a most atrocious pang; she had panted before she attacked; she had bled before she combated. In the final nights, the cries of the multitude seemed like cries of supplication toward

who once numbered seven thousand. The houses of the barbarians emerge from our ruins, and it seems that their stones are mixed up with the bones of our fathers, and that their cement is dissolved with the blood of our wound. [...] Do not forget, brothers. Do not let us perish. We, too, are Latin people [*gente*], devoted to the Latin name. [...] Who speaks of our olive and oak trees, our bushes of sage and thyme, our mildness? What does it matter if we are not Italians on Italian land [*in terra d'Italia*]? We want to rebel, to combat. We are with you, we are here for you. Here is our blood. Do not refuse it. Take it" (153–154).

47 Spackman points out the defensive form that violence assumes in Fiume, stressing that, "The perpetrators of violence—the Italian government and the protagonists of WWI—have nothing to fear from their victims at Fiume, no reprisals, no physical violence, nothing but symbolic action, sublime but sublimated violence" ("*Il verbo (e) sangue*" 222).

48 For additional context, see Renzo De Felice's commentary (533, n.5).

> a redemptive god. And then she fell silent, when her people began dying on the soil, when her people began dying on the sea, when her blood began to flow, when her best blood began to smoke before an invisible grandeur that was her promised grandeur.]

Here, blood from the *madre-patria*'s labor flows directly into the blood of her soldier-progeny, which, spilled across territory, legitimates its occupation.[49] Moreover, removed from the Abruzzi countryside, where defective multitudes soaked in bad blood threatened Giorgio Aurispa and death triumphed, here, the heroic multitude instead possesses Italy's "best blood."

While, in the passage above, *Italia* is figured as a mutilated mother, elsewhere she is presented as "enslaved and brutalized Italia," "Italia serva, abbruttita" (318). D'Annunzio draws from a topos of an anthropomorphized and gendered Italy which dates back to (at least) Dante: "Ah, slavish Italy, dwelling of grief, ship without a pilot in a great storm, not a ruler of provinces, but a whore!" (97) ("Ahi serva Italia, di dolore ostello/ nave senza nocchiere in gran tempesta/ non donna di provincie, ma bordello!") (*Purg.* VI, 76–78).[50] In his canzone "Italia mia," Petrarch figures Italy as his a "kind, mother, merciful" ("madre benigna et pia")—the point of origin for the "ancient valor in the Italian heart" ("antico valore/ ne l'italici cor") (95–96)—which Petrarch figures as contaminated by the "barbaric blood" ("barbarico sangue") (22) of foreign mercenaries, "the German rage [...] that managed to contrive a way to make this healthy body sick" (204–211) ("la tedesca rabbia/ [...] ch'al corpo sano à procurato scabbia") (39).[51] In both his autobiographical *Vita* (1803), and his scathing critique of the French occupation of Italy *Il misogallo* (1799), author and dramaturge Vittorio Alfieri (1749–1803) denounces Italy's servitude to foreigners: "viewing Italy wholly degraded from her rank as nation; and the Italians divided, weak, and enslaved, I was

49 Spackman writes of D'Annunzio's use of this figure, "[T]hree losses are figured together—the loss of the war, the loss of blood, and the loss of virginity—for we may also read the image as one of violation, of violation as a necessary though lamentable prelude to generation" (*"Il verbo (e)sangue"* 224).

50 Alberto Banti recounts that the iconography of *Italia* during the Risorgimento most often depicted her as a woman who was at times chastely clothed, at other times bare chested (underscoring her role as a nursing mother), and often either in chains (to underscore her oppression by foreigners) or armed (to underscore, instead, her heroic rebellion) (*La nazione del Risorgimento* 67).

51 Quotes are taken from Petrarca, *Rerum vulgarium fragmenta*; Petrarch, *The Canzoniere, or Rerum vulgarium fragmenta*.

ashamed of being an Italian" ("vedendo l'Italia tutta esser morta; gl'italiani, divisi, deboli, avviliti e servi; io grandamente [*sic*] mi vergognava di essere [...] Italiano") (*Vita* 65). Ugo Foscolo (1778–1827) also deploys the topos of Italy as "prostituita" by other nations and threatened by the specter of a degeneration that would place Italians on par with black slaves: "Perhaps the day will come when we, having lost our possessions, our intellect, and our voice, will be made similar to the domestic slaves of the ancient lords or traded like wretched negroes" (92). ("E verrà forse un giorno che noi perdendo le sostanze e l'intelletto e la voce, sarem fatti simili agli schiavi domestici degli antichi, o trafficati come i miseri Negri")(132). In his early patriotic hymn "To Italy" (1818), Giacomo Leopardi (1798–1837) takes up the figure of the enslaved woman:

> Or fatta inerme
> Nuda la fronte e nudo il petto mostri
> Oimè, quante ferite
> Che lividor, che sangue!
> oh qual ti veggio
> Formosissima donna! [...]
> Chi la ridusse a tale? E questo è peggio
> Che di catene ha carche ambe le braccia [...].(*Canti* ll. 6–13)

> [F]or undefended
> You show your forehead and your breast now, bare.
> What a mass of wounds, alas!
> What bruises and what blood!
> What disarray,
> My lovely lady! [...]
> Who brought her to this pass? Who dared to lay
> Upon her arms these fetters and their weight? [...].(*The Canti* 3)

From this admittedly reductive survey of the Italian canon, we can delineate the contours of this topos in order to identify how D'Annunzio reshapes it. *Italia* is, at best, a benevolent and pious mother (Petrarch) and, at worst, a prostitute, raped and enslaved (Dante, Alfieri, Foscolo, Leopardi, D'Annunzio). This gendering is, perhaps not surprisingly, operative in figuring an Italian population that is marked in various ways by blood. Petrarch warns that compromising the "ancient valor in the Italian heart" is the palpitation of another blood—that of a population that barbarizes as it invades. For

Leopardi, an enslaved nation is figured by a sullied and bloodied female body. The blood from these wounds, cautions Foscolo, can collide with the blood of race: a threat to the corporeal *Italia* reduces the status of her offspring to that of black slaves. Dismembering the topos has laid the ground for its D'Annunzian refiguring.

In an address to legionnaires aboard the *Bronzetti* as the Fiuman campaign reached its conclusion in December 1920, D'Annunzio's corporeal *Italia* is absorbed by an embodied *Patria*: "We knelt down, not you all before me, but all of us together before the *Patria* that was as present in the offering as the Lord is present in the host" (409). This rhetoric of patriotic communion both fraternal and sacred recurs throughout the speeches, and here it is figured as D'Annunzio immersed in the crowd, "heart to heart" (411). Pulsing through these devoted hearts is nothing less than explicitly racialized blood: "Allow me to listen to your blood, your good, popular blood [*sangue di popolo*], the blood of my grieving race [*razza dolente*], of my poor dear Italia" (411). The sacralization of the *Patria* and the blood spilled in order to constitute and defend it is thus effected through the "bloody mother" who gives birth to racialized offspring.

The bloody delivery that ensures the (re)birth of the nation is also figured as another kind of labor. One mutilated body (that of the bleeding mother *Italia*) thus begets another (that of the mutilated soldier). Describing "the most beautiful Italian battle of all time," when Fiuman soldiers succeeded in commandeering a Hungarian ship full of grain, D'Annunzio fuses work and battle to produce a figure for nationalist transubstantiation:

> Tutta l'aria aveva una tempra eroica. La tempravano i nembi, la tempravano la grandine e il vento. Bagnato di piogge improvvise, il grano sembrava carico di rugiada fino a mezzogiorno, fino a vespro. I fanti associati succhiavano le spighe. I carri colmi di truppe passavano su la linea del fiume inghirlandati di fronde e di fiori. I soldati cantavano, a tenzone. Le falci e le baionette, le roncole e i pugnali davano lo stesso guizzo. La battaglia ferveva come il lavoro. I contadini gridavano: "Non passano." E mietevano. I soldati gridavano: "Non passano." E combattevano. Il sangue sparso consacrava il pane rinato. Il pane rinato prometteva di rifare il sangue sparso. (253)
>
> [All the air had a heroic tone. The rain clouds tempered it, the hail and wind tempered it. Wet with sudden rains, the grain seemed laden with dew until midday, until vespers. Infantrymen sucked on ears of wheat.

> Wagons overflowing with troops passed along the river wreathed in fronds and flowers. Soldiers sung in a tenson. Sickles and bayonets, billhooks and daggers flashed in unison. The battle blazed like work. The peasants cried, "They shall not pass." And they harvested. The soldiers cried, "They shall not pass." And they fought. Shed blood consecrated the born-again bread. Born-again bread promised to restore the shed blood.]

As in Franchetti's fantasized Eritrean colony, the work of peasants and soldiers converge to produce the body and blood of the new Italian nation. In both Franchetti's and D'Annunzio's projects, this rebirth of the nation necessarily occurs outside its borders—in the contested (or "lost") lands of the (Roman and pre-World War I) past. Yet what is noteworthy about D'Annunzio's formulation is that the sacralized blood that flows in Fiume belongs not only to the "bloody mother," but also to the war dead and to the nation's mutilated heroes.[52] If military occupation is staged as work, a further, phantasmagorical link is thus drawn between the body of the mutilated soldier-hero and that of the peasant: if the latter's labor is enabled through the use of his arms (as the metonymical relationship of *braccia,* or *bracciante,* to labor suggest), the former's labor relies upon a sacrifice of them.[53]

Indeed, in D'Annunzio's Fiuman writings, a privileged figure for the heroic and mutilated infantryman is the one-armed, or altogether armless, man. Some months before the march on Fiume in September 1919, D'Annunzio launched his Fiuman campaign with the "Letter to the Dalmatians" (January 15, 1919), which Paolo Alatri has called, "the most famous imperialist

52 In 1921, the Roman veterans' organization Società Mutuo Soccorso Giuseppe Garibaldi organized what Claudio Fogu has called, "the most successful national ritual ever performed in the pre-fascist era," (328) by celebrating the transfer of the body of the Unknown Soldier from Aquileia (Veneto) to Rome's Altare della Patria. David Atkinson and Dennis Cosgrove point out that this corporeality was central to the symbolism of Rome's monument to Vittorio Emanuele II, the Vittoriano (inaugurated at the World's Fair in 1911). The Vittoriano became home to fascism's imperialist spectacle, which included ritualized parades of *mutilati* who, "proudly displayed their broken bodies as physical witness to Italian heroism" (43). On the Tomb of the Unknown Solider as an emblem of modern mourning, see Wittman, *The Tomb of the Unknown Soldier.*

53 Furthermore, the metonymical relationship of *braccia* to *bracciante* enacts a parallel linguistic dismemberment of the peasant body through synecdoche. The "whole" of the laboring peasant is rendered linguistically as a set of arms. For more on the significance of D'Annunzio's poetics of sacrifice, particularly in *Le vergini delle rocce* and his autobiographical nocturnal writings, see Pireddu, "Gabriele D'Annunzio." For more on the rhetoric of sacrifice under fascism, see Ferrari.

manifesto of the post-[World War 1] era."[54] In this letter, which announces the Fiuman campaign as it inaugurates the sacralized style that will characterize subsequent addresses, the "humble infantryman with no name" who "had a scar on his forehead, and whose arms were not visible" (7–8) is figured as an exemplary patriot, wounded and devout. The "humble infantryman" is exemplary not only in his flawed embodiment, but also as he symbolizes battles past ("He resembled the infantrymen of the Carso"), while anticipating battles future (he kneels at an altar which represents altars of the other Adriatic territories that D'Annunzio figured as *irredenti*, or unredeemed: Zara, Sebenico, Spalato, Traù, for example).

Soon thereafter, referring again to the armless infantryman, D'Annunzio insists upon his solidarity with his soldiers, claiming: "I am your equal, as the day before yesterday between an altar and a gate I felt I was the equal of the young and silent infantryman" (18). Later, he queries, "What do the true Italian people [*il vero popolo italiano*] want? With whom do they side?" One of three to step forward is "a maimed *romagnolo* [*un mutilato romagnolo*], who has only one arm" (54). The "true Italian people" are thereby represented by a truncated body, and, significantly, this mutilation involves not the loss of legs, but the loss of arms. At the thematic level, then, the renewal of the Italian race occurs first through embodiment (from the mutilated body of the mother *Italia* is born the *popolo, razza, stirpe, gente,* and/or *moltitudine*), and then through dismemberment. The figure of the one-armed *mutilato* thus mirrors the territorial dismemberment of Italy's "mutilated victory." Mutilation and sacrifice are two more terms to be added to the rhetorical constellation of loss that traverses post-Unification Italian culture's production of *vital subjects*: for D'Annunzio, lands and limbs that were lost (through "mutilation," dispossession, or sacrifice) can presumably be recovered (as we will see, modern aviation will prove crucial to this fantasy). The primary ideological fantasy of D'Annunzio's occupation of Fiume is that of territorial restitution, making Italy 'whole' again, as if it once was to begin with. His Fiuman project may thus be read as fundamentally fetishistic, insofar as a rhetoric of loss is mobilized to disavow a fundamental absence. Like the one-armed man, the Italian national body was never whole to begin with, nor were its political borders ever natural, but all the same …

54 Here, I refer to Paolo Alatri's essay, "Ideologia e politica in D'Annunzio," which conducts a brief survey of D'Annunzio's oeuvre and select criticism through static and primarily thematic conceptions of ideology and politics (Valesio, *D'Annunzio a Yale* 25).

While D'Annunzio's occupation of Fiume is initially figured as an extension of the Italian nation-state, meant to heal the territorial dismemberment represented by the *vittoria mutilata* of World War I, given the refusal of Francesco Saverio Nitti's leftist government to support the Fiuman occupation (and his movements, however hesitant, against it), he eventually frames his project in opposition to Italy in particular, and to "the West" in general:[55]

> Intanto, alla Lega delle Nazioni, a questo complotto di ladroni e di truffatori privilegiati, noi opporremo la Lega di Fiume, raccogliendo qui i rappresentanti di tutti quei popoli che oggi patiscono l'oppressione e che vedono atrocemente mutilate le fibre viventi dei loro territori nazionali. (231)
>
> [Meanwhile, in opposition to the League of Nations, that conspiracy of privileged thieves and swindlers, we form the League of Fiume, which welcomes all of the representatives of all of those peoples who today suffer oppression and who see the living fibers of their national territories brutally mutilated.]

55 It is within his turn away from "the West," represented by Rome and the League of Nations, that D'Annunzio introduces his oft-cited "League of Fiume," or "League of Oppressed Nations." Significantly, the opposition to the League of Nations is staged in defense of a biologized terrain that their politics has "mutilated." As D'Annunzio launched these critiques from Fiume, Italian Nationalist Association (ANI) founder Enrico Corradini also held "Franco-Anglo-Saxon plutocratic hegemony" responsible for the "mutilated victory" of World War I. Mauro Marsella claims that, though D'Annunizan *irredentismo* was not the primary preoccupation of the ANI, it was the only political party to unite in support of Adriatic expansion, encouraging Orlando and Sonnino's acceptance of the Pact of London (which did not, however, include Fiume). Furthermore, during a speech delivered for the inauguration of the formation of the ANI in 1910, Corradini adapted revolutionary syndicalist Arturo Labriola's figure of the "proletarian nation" to describe Italy's challenged relationship to other western European powers. Giovanni Pascoli would also deploy this figure in his support for the Italian invasion of Libya in 1911 with his famous address, *La grande proletaria si è mossa* (The Great Proletarian, She Has Risen!). For Nicola Labanca, Leopoldo Franchetti's solutions to the questions of emigration and colonization in the liberal period (see Chapter One) were part of a larger public discourse that presented Italian colonialism as based more upon the peaceful exercise of labor than upon the violent march of capital. Labanca draws a direct line of continuity between this discourse and Corradini's and Pascoli's subsequent cries for a "proletarian" colonization of Libya. See Labanca, *Oltemare* 371–372 and Chapter Four of this volume.

The loss of territory is not simply an unjust geo-political reshuffling at the hands of the powers that be, it is the "brutal [mutilation]" of the "living fibers" (or *vital subjects*?) of the nation. D'Annunzio draws an explicit line of distinction between "fiumani" and "italiani," insisting: "We do not want to be Italian" (242). Increasingly throughout the speeches, he figures Fiume as the site for a reversal of the Italian degeneracy represented by an inept Rome. On August 5, 1920, he addresses the "Legionaries of Fiume, the only force of a non-bastardized Italy" (300). His figuring of the Italy outside Fiume as "bastardized" thus solidifies Fiume as the terrain, presumably not marked by a generative illegitimacy, where a connection between the sacralized blood of heroes and the regeneration of the Italian race may be forged.

D'Annunzio's writings therefore also require the grafting of extant Italian nationalist vocabularies (the "genius of the race" and famed Italian artistic, literary, and political genealogies) onto a Fiuman one, which defines itself in part through an opposition: we are simultaneously you and greater than you, because we are not you, goes the logic ("a cry that is doubly Italian" sounds alongside the declaration "we do not want to be Italian" 290; 242). Likewise, metaphorical and rhetorical modes that are central to early Italian colonial discourse are woven into D'Annunzio's call for racialized territorial appropriation in the Adriatic, one of whose constituent parts is a rhetoric of anti-colonialism. In forging this new and improved bio-territory the rewriting of another narrative is required—arguably the most quintessentially Italian narrative of loss insofar as it is profoundly embedded in the literary and political writings of *fin-de-siècle* Italy—that of Italian emigration. Between 1914 and 1919, D'Annunzio had published several messages "To the Italians of the United States," calling for the support of his Italian brothers in World War I and later for their support of his Fiuman campaign. He opens one of his pro-Fiuman messages with, "Brothers, do you remember? There was no longer supposed to be an ocean between this dolorous *Patria* and the great filial colony. There was no longer supposed to be the Atlantic between the *Patria* who bled from the blow of the enemy and the one that was made overabundant with your work."[56] In order to further substantiate these claims, I want to turn to a lesser-known speech, "L'Ala d'Italia è liberata," which D'Annunzio delivered to aviators at the airport of Centocelle (Rome) in July 1919, roughly two months before his arrival at Fiume.

56 "Agli Italiani degli Stati Uniti," in D'Annunzio, *Prose di ricerca* Vol. I 991. See also Bertazzoli. For a brief account of D'Annunzio's reception by Italians in the United States, see Scarlini.

From Arms to Wings

Historians and biographers have described D'Annunzio's fascination with aviation as part of a modernizing aesthetic in both his literary and autobiographical projects.[57] Michael Ledeen suggests that his love for aeronautical adventure initially stood to distract D'Annunzio from taking the lead in Fiume. "As the forces of sedition attempted to figure out their best strategy in the Adriatic," Ledeen writes, "D'Annunzio toyed with a variety of lofty projects. Should he join with these men and 'liberate' Fiume and/or Dalmatia? Or should he embark upon a long-discussed and highly dramatic exploit—a plane trip to Tokyo?" (55–56).[58] Literary scholar Annamaria Andreoli reveals another of D'Annunzio's linguistic novelties in his proposed substitution of *aeroplano* with *velivolo* (*D'Annunzio* 150).[59]

D'Annunzio's speech to aviators at Centocelle in July 1919 was a kind of homecoming: he had been there in the spring of 1909, when Wilbur Wright was invited to help establish an Italian school of aviation. Among other distinguished pupils, Wright trained Italy's first pilot, Lieutenant Mario Calderara, while the poet eagerly looked on. Later that year, in September, D'Annunzio attended the Air Show at Brescia to see Calderara again, along with famed international pilots Glenn Curtiss and Louis Blériot.[60] Other

57 Just as D'Annunzio's occupation of Fiume is read as a turning point in the modernity of political spectacle, D'Annunzio's aerial exploits are celebrated as a constitutive part of the modernization of Italian aviation. Though the Italian Air Force was not officially formed until March 1923, historian of Italian aviation Giuglio Lazzati places D'Annunzio's famous "flight over Vienna" in 1918 within the origins of the modern Italian Airforce (Aeronautica Militare Italiana) (10). In fact, according to historian Giuliano Colliva, the flight was made with the latest in Italian aeronautical innovation at the time—seven fighter biplanes (Ansaldo SVAs), six of which were one-seaters carrying only a pilot and the seventh of which was a two-seater carrying a pilot and, as Colliva puts it, D'Annunzio as an "observer."

58 On the poetics of aviation in Marinetti's "Manifesto tecnico della letteratura futurista" (1912), see J. Schnapp, "Propeller Talk." From *La Nave* (1907) to *Forse che sì forse che no* (1910), we may read a shift from the technology of marine conquest to an aerial one. As Croce christens D'Annunzio the "spiritual father" of reactionary Italian imperialism and nationalism, he cites *La Nave* and *Forse che sì forse che no* as examples of his "imperialistic" linguistic style (*Storia d'Italia* 244–245).

59 Peter Demetz disputes D'Annunzio's coining of the term, citing the dictionaries of Algarotti (1764) and Florio (1598). The Latin *velivolu(m)* referred to the sea plowed by ships ("D'Annunzio Poet and Aviator" 157).

60 For more on these and other pilots present, see Demetz, "Aviators at Brescia."

distinguished attendees included King Vittorio Emmanuele III, Giacomo Puccini, F.T. Marinetti, and Franz Kafka, who was covering the show for the Prague daily newspaper *Deutsche Zeitung Bohemia*. D'Annunzio was busy working on his "romanzo dell'ala," or aviation novel *Forse che sì forse che no* (to be published in 1910), which would continue to explore the figure of the Nietzschean *Übermensch* or Superuomo—as D'Annunzio had in Claudio Cantelmo of *Le vergini delle rocce*, Giorgio Aurispa of *Trionfo della morte*, and Stelio Effrena of *Il fuoco*—this time through the character of Paolo Tarsis, whose force and vitality were expressed through his mastery of modern technology, and aviation in particular (Syrimis 148). D'Annunzio arrived at Brescia on September 10 in order to spend a few days tracking the technical procedures and vocabularies of professional aviators and mechanics, and most importantly, to take his first flight in a biplane.[61] Peter Demetz describes some of the details the writer jotted down while touring the hangars at Brescia:

> [T]he mechanics' folded beds, their coarse boots dirtied by the *brughiera* soil, their sweaty shirts, iron wires hanging from a nail on a wall [...] A few notes described how a pilot, or a mechanic, fills the gas tank, first measuring what he needs with a dipstick and then pouring gasoline through a yellow cloth filter from a canister painted white. ("D'Annunzio: Poet and Aviator" 155)[62]

Though the poet's first flight was ushered in with great fanfare and almost immediately achieved the status of legend, in reality, it was a relatively brief event. After unsuccessfully soliciting Louis Blériot for a ride, a friend of D'Annunzio's twisted the arm of the record-breaking American pilot Glenn Curtiss, who eventually acquiesced. Curtiss and D'Annunzio were airborne for a matter of minutes before somewhat abruptly touching back down. The following day D'Annunzio would enjoy a slightly more sustained trip alongside Calderara. Despite the brevity of his time spent aboard aircraft at

61 Annamaria Andreoli notes that *Forse che sì forse che no* was initially to include the subtitle *Romanzo dell'ala* (D'Annunzio, *Prose di romanzi* 2 1316).

62 Demetz also points out that D'Annunzio was at least as captivated by the landscape as he was by technical details, as he "even add[ed] a little pencil sketch showing how the struts and braces of an airplane, *la nervatura delle ali*, form a distinct pattern of squares when seen against the sun." Demetz remarks, "He simply cannot resist aestheticizing technical matter" (156).

Brescia, D'Annunzio related his experiences to journalist Luigi Barzini with characteristic exuberance:

> Una sensazione divina! [...] Vorrei essere aviatore! [...] Vorrei poter salire a centinaia di metri nello spazio! [...] Oh, io abbandonerei tutto, tutto per dedicarmi all'aviazione! Invidio questi uomini che hanno fatto del volo lo scopo della loro vita. (Andreoli, *Il viviere inimitabile* 448–449)
>
> [A divine sensation! [...] I wish I were an aviator! [...] I would like to be able to soar hundreds of meters into space! [...] Oh, I would abandon everything, everything to dedicate myself to aviation! I envy these men who have made flying their life's purpose.]

D'Annunzio would publish his successful novel *Forse che sì forse che no*, fusing a heroic history of Italian aviation with a narrative of erotic passion, the following year, drawing in large part from his fleeting firsthand experience with flight, and from the meticulous note taking he had done at Brescia. The publication of his technological novel prompted an invitation to deliver public lectures on the theme of "The Mastery of the Heavens" (*Il dominio dei cieli*) in several major northern Italian cities, from Milan and Turin to Venice and Bologna (Woodhouse, *Gabriele D'Annunzio* 244).[63] Though he missed an opportunity in October 1911 to fly with the Italian air force over Tripoli (an opportunity that F.T. Marinetti would instead seize) *il poeta-aviatore* would fly again between August 1915 and June 1916 (at the age of fifty-two), dropping anti-Austrian pamphlets over Trento and Trieste.[64] The second plane was attacked by Austrian anti-aircraft fire, causing a grave injury to D'Annunzio's

63 The lectures are published in Saverio Laredo de Mendoza, *Gabriele d'Annunzio aviatore di guerra* (Milan, Editoriale Italiana, 1931) 69–85 (cited in Demetz, *The Airshow at Brescia* 240).

64 D'Annunzio would instead commemorate the invasion with his "Canzoni della geste d'Oltremare," in which he praised the "glory of the race" [*razza*] of Aeneas in its effort to conquer Libya and "widen [Italy's] skies" ("La canzone del sangue," *Merope* 29). The *Canzoni* were published in the *Corriere della Sera* between October 1911 and January 1912, and are now contained in *Merope*, book four of *Laudi del cielo, del mare, della terra, e degli eroi*. Peter Demetz draws a direct line of continuity between D'Annunzio's missed opportunity to participate in the historic flight over Tripoli during the Italian invasion of Tripolitania and Cyrenaica beginning in October 1911 and his participation in the early Italian film industry in Turin (and specifically on Giovanni Pastrone's early epic *Cabiria*). See also Chapter Four of the present study.

right eye and nearly blinding him. In his convalescence, he began to pen, on "more than ten thousand" strips of paper, an autobiographical prose poem, *Notturno*, composed in 1916, published in 1921 (Andreoli, *Il vivere inimitabile* 537).[65] The diary of his "blind immobility" became the supreme expression of a poet who sought the heroic life on the front lines, and the fragmented conditions of its composition along with its telegrammatic style mirrors D'Annunzio's own corporeal mutilation.[66] Nor did his injury signal the end of his desire to fly—on August 9, 1918, D'Annunzio was a passenger on a biplane that flew alongside eight others from the San Pelagio airfield (near Padua), writes Giorgio Evangelisti, to drop hundreds of thousands of pamphlets on the city of Vienna, celebrating the liberated nations of Austria-Hungary and urging the Austrians to cut their ties with Germany.

Like his novel *Forse che sì forse che no*, D'Annunzio's July 1919 address to the aviators at Centocelle, "L'Ala d'Italia è liberata," certainly champions modernist innovation. In D'Annunzio's earlier aviation novel, heterosexual relationships between antagonistic male and female characters are mediated by technological objects. Michael Syrimis argues that this constitutes a departure from D'Annunzio's *fin-de-siècle* decadent novels insofar as in those novels, women are often associated with "soft-textured fabrics, such as the animal fur, the divan, the cushions [...], or nature that is still organic" (148). In *Forse che sì forse che no*, mechanical modernity takes center stage as the means through which characterization is carried out. Paolo longs for his lover Isabella to be as manageable as machinery, and describes her instead as machine-like (Syrimis 148). D'Annunzio carries out a similar association in his speech to aviators as he fuses the embodiment of racialized genius in the people (in this oration figured variously as *stirpe*, *razza*, *popolo*, and

65 Nicoletta Pireddu argues that in *Notturno*, D'Annunzio posits death and the sacrifice of both the work of art and the author as the ultimate gift that escapes utilitarian exchange. See Pireddu, "Gabriele D'Annunzio" 178–179.

66 Furthermore, as Paolo Valesio argues, it is in *Notturno* that D'Annunzio turns to the figure of the *Miles patiens*, "the Italian soldier seen as a victim, as a witness in the original and strong sense of the word (thus, a martyr)." Valesio claims that by the end of the Great War, D'Annunzio "had intuited something that fascism was never able to comprehend [...]. What the poet sensed—with that blend of short-term ingenuousness and historical farsightedness that often characterizes great poets—was that Italy had exhausted itself in the war effort [and] its imperial role was definitively lost." *Notturno* thus represents "the literature (or better, the poetry) of this political intuition. [...] He begins to fashion (over and beyond any polemics about a "mutilated Victory") a discursive vision of Italy as a martyr or victim" (*Gabriele D'Annunzio* 119; 147).

cittadini) to aeronautical innovation; this gesture, according to the logic of the text, stands to mend both the territorial and corporeal dismemberment that D'Annunzio's Fiuman project seeks to remedy. The mutilated and fetishized body of the soldier-patriot who occupies a central place in D'Annunzio's Fiuman imaginary (a figure for the "mutilated," or incomplete, Italian peninsula) stands to regenerate, by sprouting wings and amassing territory:

> L'istinto icarico, l'istinto umano del volo, che già travagliava l'inquietudine del Vinci e si rivelava nei disegni esatti e nei congegni reconditi, s'è approfondito e non ha preso vigore in nessuna stirpe come la nostra. La volontà dell'aquila romana, che precedeva per tutta la terra la marcia cadenzata dei legionari, sembra rinascere nei nostri giovani stormi. Essa non è più una insegna di milizie pedestri; è una crescente rapidità di conquistatori aquilei. Pareva che fra vent'anni, fra trent'anni, fra cinquanta dovessimo avere una generazione in cui l'ansia del volo fosse già trasmessa come un retaggio. Per una di quelle anticipazioni che sono i prodigi del nostro genio, l'ansia s'è risvegliata nella gioventù presente. Il giovine corpo della nostra aviazione ha l'istinto spontaneo dei grandi migratori alati: il bisogno di migrare, di peregrinare, di spandersi nei quattro venti. Più perfettamente che le due nature del centauro, l'uomo e il velivolo fanno una sola forza veloce. (92)[67]
>
> [The Icarian instinct, the human instinct for flight, which aggravated the restlessness of da Vinci and revealed itself to him in exact designs and hidden mechanisms, has penetrated and taken vigor in no other race [*stirpe*] as in ours. The will of the Roman eagle, which preceded the rhythmic march of the legionaries across the Earth, seems reborn in our young flocks. It is no longer a badge worn by pedestrian troops; it is a growing swiftness of aquiline conquerors. It seemed that within twenty, thirty, fifty years, we were bound to have a generation in which the longing for flight was transmitted as a legacy. Thanks to one of those forecasts that are the wonder of our genius, that longing has been reawakened among the young people here today. The young of our aviation corps have the spontaneous instinct of great winged migrants: the need to migrate,

67 D'Annunzio's figuring of the military masses as "stormi" also marks a transition from marine to aeronautical rhetoric: no longer the turbulent aqueous mass of D'Annunzio's earlier novels, the masses are redeemed and projected skyward as heroic "flocks."

> to roam, to spread out in the four winds. More perfectly than the dual nature of the centaur, man and aircraft form a single, swift force.]

Here, D'Annunzio returns to the figure of Icarus, whom he had first praised in his 1903 poem "L'ala sul mare" (*Alcyone*) (and therefore just as the Wright brothers were preparing for their historic first heavier-than-air human flight), as he sketches a uniquely Italian genealogy of flight—a direct line of descent from Icarus to da Vinci to the modern aviation corps present at Centocelle, who are armed with the "will of the Roman eagle." Aeronautical prowess is genetic, handed down from generation to generation, increasing in vigor until man and plane become one. Indeed, in this speech, which places Italian aviation at the center of D'Annunzio's irredentist project, the "mutilated victory" is figured as the result of "the geese's wretched hatred for eagles." D'Annunzio recalls an address to aviators on the eve of his famous flight over Vienna in August 1918: "The sadness of constrained idleness, the tedium of waiting in vain, the anguish of the mutilated and agonizing victory, and the geese's wretched hatred for eagles oppressed us" (99). Technological mastery thus stands to restore Roman primacy, as the ancient eagle resurfaces, metallicized, motorized, and triumphant.[68]

This speech is also an important locus of another convergence: within this address, as I have mentioned, the stakes for the recast relationship between blood and race that I have been tracing are made explicit, as aeronautical innovation is a "divine instinct," representative of "the genius of the race [*razza*]," which stands to enable ever-more territorial acquisition (91). For Jeffrey Schnapp, constitutive of the "aviation craze" in Italy, as elsewhere, was a link between aerial and territorial conquest. Schnapp highlights a connection between the "airspace" of F.T. Marinetti's "Manifesto tecnico della letteratura futurista" (Technical Manifesto of Futurist Literature, 1912) and colonial expansion, which posited the air, like the African continent before it, as, "humankind's final frontier: a last remaining source of mystery, miracles, novelty, and the unknown [...] Something to be preserved, therefore, in its 'savage' natural state even as it is conquered and absorbed

68 While D'Annunzio (along with Italian colonialists past and future) wields Roman primacy to legitimate his occupation, American President Wilson was not so convinced. Burgwyn recounts that when British Prime Minister Lloyd George suggested that one solution to the Adriatic situation would be to offer Italy territories in Asia Minor, claiming, "The Romans were very good governors of colonies," Wilson quipped, "Unfortunately, the modern Italians are not the Romans" (cited in Burgwyn 277).

within earthbound laws and institutions" ("Propeller Talk" 169–170). In his fetishistic celebration of aviation, D'Annunzio produces a genealogy of Italian migration that posits a direct line of continuity between "medieval" exploration and modern colonization:

> Il popolo italiano fu sempre il più sagace dei migratori. [...] Nell'Evo medio, nel Rinascimento, nell'età più tarda, l'uomo italiano fu re in tutti i mari, fu signore in tutte le terre, sino agli ultimi orizzonti, sino agli estremi confini. Quell'Africa e quell'Asia, che oggi gli sono contese dalla perfida avarizia altrui, furono sempre alla mercè de'suoi ardiri. Ma non importa che gli sieno contese. "*Teneo te, Africa*" è una parola romana da rendere italica. Chi può reprimere in noi questa volontà, questa attitudine, questa tradizione? Non era possibile quando non avevamo se non la chinea ambiante e il cavallo di San Francesco, o la vela e il remo? Sarà possibile oggi che il nostro vecchio istinto migratorio ha messo le ali, le sue giovani ali? (96)[69]
>
> [The Italian people were always the most shrewd migrants. [...] In the middle ages, during the Renaissance and after, Italian man was king of all seas, he was lord of all lands, to the final frontier, the farthest reaches. The Africa and Asia that are sought after today by the perfidious avarice of others were always at the mercy of his boldness. But it doesn't matter that they are sought after. "*Teneo te, Africa*" is a Roman promise to be rendered Italic. Who can repress in us this will, this habit, this tradition? Was it not possible when we had only the horse of Saint Francis, or the sail and the oar? Will it be possible now that our ancient migratory instinct has sprouted its youthful wings?]

The Italian history of migration is thus rewritten as one of naturalized heroic instinct, rather than one of economic necessity ("Italian man was king of all seas" and "lord of all lands"), and it is this heroic instinct, as well as an oft-cited Roman primacy, that justifies colonial appropriation.[70] The Fiumans

69 See also D'Annunzio, "Più che l'amore" 1112.

70 The specter of Italian emigration (as a result of hunger and dispossession) also haunts the text as it is folded into the intersecting rhetorics of sacrifice and oppressive injustice leveled against Italy by France, Great Britain, and the United States: if the terrain of war is figured as blood-soaked, D'Annunzio anthropomorphizes it in order to figure its inhabitants as "insatiable." These three powers represent, "the five-meal-a-day people"

hailed by D'Annunzio's address are "oppressed" in part because they are shut out by the "perfidious avarice of others" in the scramble for Asia and Africa. D'Annunzio goes on to quote a colonial general, who anchors D'Annunzio's aeronautical vision in the biological territory of Friuli:

> Per tutte le vie dell'altura e della pianura era la stessa fecondità. Ad ogni borgo, a ogni villaggio a ogni casale i bambini robusti e coloriti erano tanti che pareva schizzassero dalle ruote della mia automobile, come schizzano le gocciole quando s'entra in un guado. Che soda materia umana! Chi può dubitare delle sorti di questa nostra razza inesauribile e incoercibile? Dimmelo. (96)

> [Along all the roads of the highlands and plains was the same fecundity. In every district, in every village and hamlet, robust and florid children abounded so that they seemed to splash from under the wheels of my automobile like drops of water when one crosses a ford. What solid human material! Who can doubt the fate of this inexhaustible and incoercible race of ours? Do tell.]

As in Franchetti's Eritrean fantasy, fertile lands make for bodies both invincible and overflowing. D'Annunzio continues, in his own words, sending these robust representatives of the race toward aerial conquest:

> Di quella buona materia umana una parte rimarrà attaccata alla sua terra; ma una parte s'involerà verso l'avventura e la conquista, una parte metterà le ali, intraprenderà le mille vie azzurre, andrà al di là dai mari dai deserti dalle montagne senza più temere l'ostacolo. (96)

writes D'Annunzio, who "are hungry": "Once it was satiated and fattened with victims, the land seems to have transmitted its hunger to peoples [*genti*]. And, if the land is satiated, man seems insatiable" (*La penultima ventura* 12–13). While the Allies are hungry for blood and territory, the dispossessed Italians hailed by D'Annunzio are pious in their sacrifice: despite their hunger, they retaliate by shedding blood. In an address on the occasion of his acceptance to lead the Fiuman expedition, entitled "Italy or Death," D'Annunzio urges: "Italy knows hunger, but it does not know dishonor" (105). Denouncing the mistreatment of Italian war prisoners, D'Annunzio describes one group: "a muddy and panting heard of defeated men crossed the city, thrust forward by the goad of the bayonet and the butt of the rifle of the Croatian cops. The citizens cried, and drank their tears in silence; and they struggled to help their barefoot, ragged, seminude brothers who were devoured by fever and hunger, who were alive only in their supplicating eyes" (107).

> [Of all that good human material, some will remain attached to their land, but others will take flight toward adventure and conquest; they will sprout wings, embark along blue paths, go beyond seas deserts mountains [*sic*] without fearing obstacles.]

The call for an alliance with the "oppressed peoples" that follows is thus enunciated through a rhetoric of racialized primacy, and one that requires figuring Italian migration as an expression of natural genius. Yet the memory of an alternative narrative of migration haunts this formulation, as D'Annunzio envisions a triumphant flight, "to our brothers in Brazil and in Argentina" (94). D'Annunzio thus uses a poetics of aviation to transcend both the rhetorical obstacles presented by this less-than-heroic history of Italian migration, and to draw a line of continuity between World War I, the occupations of the Adriatic coast, and a longed-for Italian Africa:

> Come campano nei ritrovi solitari di Tripoli e di Bengasi le nostre belle squadriglie? Oziano ancora? Bisognano gli inseguimenti e i combattimenti nel cielo del Piave e del Grappa? Perché gli aquilotti restano prigionieri nelle gabbie roventi? [...] Fate che vadano a cercare i nuovi cittadini d'Italia nei villaggi e negli attendamenti più remoti. Fate che annientino il deserto di sabbia, fate che annientino il deserto di sale. Fate che fra Tripoli e Murzucco, fra Tripoli e Tummo la via gialla si muti in via cerulea, e che l'oasi di Cufra veda giungere nel vento le ali tricolori come le bandiere tese della nuova Patria. Congiungete la Tripolitania all'Eritrea, la Cirenaica alla Somalia. I vittoriosi del cielo carsico, del cielo veneto, del cielo istriano mandateli a consolare i morti di Adua. (98)

> [How are our beautiful squadrons getting by in those solitary outposts in Tripoli and Benghazi? Do they need aerial pursuits and battles over the Piave River and Monte Grappa? Why are the eaglets kept prisoner in scorching cages? [...] Make them look for Italy's new citizens in the most remote villages and encampments. Make them obliterate the sandy desert, the salty desert. Make it so that between Tripoli and Murzuq, between Tripoli and Tumu, the yellow road transforms itself into a sky-blue road, that the oasis at Kufra sees the tri-colored wings arrive through the air like flags hoisted to honor the new *Patria*. Connect Tripolitania to Eritrea, Cyrenaica to Somalia. Send the victors of the sky over the Carso battlefield, of the Venetian sky, of the Istrian sky to console the soldiers who died at the battle of Adwa.]

Thus the regeneration of a naturalized population at Fiume, ensured through the spilling of sacralized blood and corporeal mutilation, is tethered to both a symbolic and material terrain of aerial conquest. Within D'Annunzio's Fiuman writings, territorial mutilation is both mirrored and remedied through corporeal loss—laboring arms are ideologically and thematically fused with amputated ones to ensure that the work of occupation is carried out. And from these truncated limbs, the texts suggest, wings may sprout, enabling a cerulean suturing of national territories past, present, and future. The wounds of the Italian army's early colonial defeat at Adwa are healed by the aerial combat of Libyan conquest and World War I. Italy's geographically disparate African colonies—Tripolitania, Cyrenaica, Eritrea, Somalia, and Ethiopia—are fused together not through a motorway or railway, but a "sky-blue road." At Fiume, the "mutilated victory" of the geo-political Italian peninsula is first projected onto the mutilated body of the "bloody mother," who gives birth to a soldier-patriot both racialized and truncated. Aerial conquest stands to restore wings to this second mutilated body, thereby ensuring a regeneration of the first: from the "madre sanguinosa" she may become the "Vittoria alata."[71]

From Song to Prayer: The Mutilation of Victory

Having established this Fiuman trajectory, D'Annunzio's earlier verse contains figures of this Victory both before and at the scene of her mutilation. In the passage from the "vittoria alata" (winged victory) to the "vittoria mutilata" D'Annunzio announces a thematic shift from marine to aerial conquest, which parallels the shift from a Hellenic rhetoric to the Christological one that dominates his Fiuman writings.

Well over a decade before Italy's claim to the Adriatic territories was compromised in the wake of World War I, and a decade before the onset of D'Annunzio's aeronautical fever, his hymn "Canto augurale per la nazione

71 The figure of the one-armed or altogether armless soldier-patriot through which this "winged victory" is enacted recalls another famous D'Annunzian truncation: that of Silvia Settala's hands in his 1898 play *La gioconda*. Barbara Spackman has argued that Silvia's mutilation is a kind of female castration, as her beauty is figured as a "wingless Victory." Spackman cites Armand Caraccio's equation of Silvia's wingless victory with the "mutilated victory" evoked at Fiume. See: Armand Caraccio, *D'Annunzio, dramaturge* (Paris: Presses universitaires de France, 1951) (cited in Spackman, *Decadent Genealogies* 199, n. 41).

eletta" ("Augural *Canto* for the Chosen Nation," 1889) celebrates a nation in flight:

> Un'aquila sublime apparì nella luce, d'ignota
> stirpe titania, bianca
> le penne. Ed ecco splendere un peplo, ondeggiare una chioma ...
>
> Non era la Vittoria, l'amore d'Atene e di Roma,
> la Nike, la vergine santa?
> Italia! Italia!
>
> La volante passò. Non le spade, non gli archi, non l'aste,
> ma le glebe infinite.
> Spandeasi nella luce il rombo dell'ali sue vaste
> e bianche, come quando l'udìa trascorrendo il peltàste
> su 'l sangue ed immoto l'oplite.
> Italia! Italia! [72]
>
> [A sublime eagle appeared in the light, born of nameless titans, white of feathers. And suddenly a peplum shining, hair flowing. It was Victory, the beloved of Athens and Rome, the Nike, the holy virgin. Italia! Italia! The winged goddess flew over. Not swords, bows, or arrows but endless fields. The roar of her vast white wings spread out in the light, like when the peltast and the motionless hoplite sensed her passing over in their blood. Italia! Italia!][73]

A brilliant white Victory with gleaming robes and thundering wings sails over Hellenic soldiers.[74] The *canto* goes on to describe Victory's flight across fertile agricultural lands manned by workers and her arrival at an armed port, with ships bound for the "dominion of the world" through a great battle over the "Latin sea." "Oh ever-reborn, oh flower of all races [*stirpi*], aroma of all the

72 The *canto* was originally published in the review *Nuova Antologia* on November 16, 1899. It later became the conclusion to *Elettra,* the second book of *Laudi del cielo, del mare, della terra e degli eroi* (D'Annunzio, *Versi d'amore e di gloria,* Vol. II, 407–409).
73 My thanks to Gianluca Rizzo for his assistance with this and subsequent translations of D'Annunzio's verse.
74 *Peltàste* and *oplite* refer to light and heavy Hellenic infantry respectively (*D'Annunzio, Versi d'amore e di gloria,* Vol. II 1142, n. 19).

Earth. Italia, Italia, sacred to the new Dawn, with the plough and the ship's bow" ("O Semprerinascente, o fiore di tutte le stirpi/ aroma di tutta la terra/ Italia, Italia/ sacra alla nuova Aurora/ con l'aratro e la prora!") concludes the canticle, thus returning to its opening call and fulfilling what commentator Annamaria Andreoli explains is "the semantic intention of the *canto*: joining two emblems—marine and rural—in the consecration of *Vittoria*" (D'Annunzio, *Versi d'amore e di gloria*, Vol. II 1143, nn. 55–56).[75] D'Annunzio thereby situates an Italian Victory in the symbolic union of agricultural (*aratro*) and naval (*prora*) force. "Il Fato d'Italia" ("The Fate of Italy"), he had written earlier that year, will take the form of "that ancient Victory hoisted upon a ship's bow shaped like a plowshare: to signify that the future greatness of the race (*stirpe*) will emerge only from the furrow dug deep into the earth and the fierce furrow of the sea" ("quella Vittoria antica alzata su una prora che ha la forma d'un vomere: a significare che la futura grandezza della stirpe verrà dal solco profondato nella terra e dal solco fervente nel mare").[76] The speed of the warship thereby forcefully furrows the turbulent sea, rendering it ripe for conquest, just as the plow furrows the land in preparation for its insemination. At Fiume this marine wake is recast as an aerial slipstream.

By "La preghiera di Sernaglia" ("The Prayer of Sernaglia") of 1918, the site of Victory's mutilation, Hellenic legitimacy—linked as it is in the "Canto augurale per la nazione eletta" to naval conquest—has been replaced with the Christological rhetoric that will explode across the pages of the Fiuman writings (D'Annunzio, *Prose di ricerca* 593–599). In the earlier "Canto," Victory is draped in a Greek *peplos*, surveying Hellenic legions and agricultural heroes. The symbolic fusion of battle and agricultural work at Fiume is thereby rehearsed in this "Canto." Yet the "Preghiera," setting as it does the rhetorical stage for Fiume, reinscribes the ideologemes of "work" and "battle," "sweat" and "blood" within a Christological vocabulary of mutilation and sacrifice.

The prayer opens with a rendition of the Passion in which the details of

75 This final call ("O Semprerinascente…"), which begins, "Così veda tu un giorno il mare latino coprirsi/ di strage alla tua guerra/ e per le tue corone piegarsi i tuoi lauri e i tuoi mirti," will serve as the epigraph to D'Annunzio's "Le canzoni della geste d'Oltremare" of 1911–12 (now contained in *Merope*, the fourth book of the *Laudi*), written in celebration of the Libyan War. Furthermore, the "aurora" will be evoked in the "La preghiera di Sernaglia": "Quel che in Dio fu detto è ridetto: 'Guardia, che hai tu veduto dopo la notte? Guardia, che hai tu veduto dopo la notte?' L'aurora! L'aurora!" D'Annunzio, *Prose di ricerca* 598.

76 D'Annunzio, *Taccuini* (Milan: Mondadori, 1965) XXIX, 333 (cited in D'Annunzio, *Versi d'amore e di gloria*, Vol. II 1143, nn. 55–56).

corporeal mutilation take center stage: the garden of Gethsemane is inhabited by one with a "beastly yoke around his neck" ("gogio bestiale sul collo"), a "stump that gushes blood" ("moncherino grondante"). Christ's suffering body is traversed in order to arrive at the next section of the poem, which turns to a battle between "us" and "them": "What is left in the world that they have not broken or corrupted? Their breath is more pestilent than a vulture's vomit" ("Che mai resta nel mondo, ch'essi non abbiano guastato e corrotto? Più pestilente è il lor fiato che il vomito dell'avvoltoio"). The song continues:

> Hanno arso i duomi di Dio dove battezzammo i nostri nati, portammo le nostre bare, prostrammo il nostro cuor tristo.
> Hanno abbattuto i nostri altari, fonduto le nostre campane, contaminato le nostre reliquie, maculato le specie di Cristo.
> Lordato hanno le nostre case, scoperchiato i nostri sepolcri, sterilito ogni solco, divelto ogni erba e ogni fusto, disperso i semi, corrotto le fonti, percosso i vecchi, forzato le donne, fatto monco ogni fanciullo robusto.
>
> [They have torched God's cathedrals, where we baptized our newborns, carried our coffins, laid our sad hearts. They have razed our altars, melted our church bells, contaminated our relics, stained Christ's species. They have soiled our homes, uncovered our sepulchers, sterilized every furrow, uprooted every grass and stem, dispersed the seeds, contaminated the wells, beaten our elders, raped our women, and maimed every robust young man.]

Outsiders have desacralized and sterilized the terrain, raped its women, and maimed its otherwise hardy boys. Yet, rather than establishing the conditions for resignation or defeat, D'Annunzio's Christ rises to fight:

> Ma Colui che già pianse per Lazaro, Colui che sopra Gerusalemme già pianse, Colui che già pianse nell'Orto, oggi piangere non può sopra il mondo.
> Non piange più; combatte. Non ha il capo chino su l'omero scarno, né inchiodate le palme all'infamia, né i piedi trafitti. [...]
> Ma lo vede ogni fante, simile a sé, con l'elmetto del fante, con le uose del fante, col sudore e col sangue del fante, allato allato.[77]

77 If we allow the homonymic relationship to speak, this Christ-like soldier may indeed be "winged" ("al[l]ato, al[l]ato").

> [But he who once cried for Lazarus, He who once cried over Jerusalem, He who once cried in the Garden, can no longer cry for the world. He no longer cries; he fights. He does not rest his head on an emaciated humerus, his palms are not nailed in infamy, nor are his feet pierced. [...] But every foot solider sees him, like them, with the soldier's helmet, the soldier's gaiters, the soldier's sweat and blood, nearby, nearby.]

Christ and soldier are thus symbolically fused, the mutilation of the former legitimating that of the latter. It is this mutilation that is a call to battle: "Oh fetters, thorns, flagellation, renunciation and shame, pus and blood, oh passion of Christ and of the world, oh victory beyond death!" ("O vincoli, o spine, o flagella, rinnegamento e vergogna, soma e ambascia, sete e fame, sanie e sangue, o passione di Cristo e del mondo, o vittoria al di là dalla morte!"). The prayer concludes with the famous cry denouncing Italy's "mutilated victory": "Our Victory, you will not be mutilated. No one can shatter your kneecaps or clip your wings" ("Vittoria nostra, non sarai mutilata. Nessuno può frangerti i ginocchi né tarparti le penne"), and announces a winged victory that encompasses not only land and sea, but a limitless sky: "The skies are less vast than your wings" ("I cieli sono men vasti delle tue ali.") The "Preghiera" thus forecasts Fiuman rhetoric as it reinscribes the terms that envelop Victory in the earlier "Canto," placing them within an economy of mutilation, sacrifice, and, in the final instance, boundless aerial conquest.

At Fiume, the "mutilated" lands of the Adriatic found a bodily correlative in the one-armed *mutilato*, or war-wounded man. For D'Annunzio, aviation worked to heal truncated limbs both territorial and corporeal—by sprouting wings, soldier-heroes could amass territories that had been wrested from the nation-state. D'Annunzio's aeronautical fever ought thus to be linked to his rhetorical equation of agricultural work with war and migration with conquest. As fascist propaganda would celebrate with ever-increasing zeal—shrewdly overturning Isaiah's prophecy that God's people "shall beat their swords into plowshares and their spears into pruning hooks" (2:4)—in D'Annunzian rhetoric at Fiume, the tools of the peasant laborer become the arms (in both senses) of war. Likewise, migration out of desperation (which Leopoldo Franchetti, for example, sought to redress two decades prior through his Eritrean project) is refigured as technological innovation and heroic colonization.

> Who can repress in us this will, this habit, this tradition [of migration]? Was it not possible when we had only the horse of Saint Francis, or the sail

> and the oar? Will it be possible now that our ancient migratory instinct has sprouted its youthful wings? (*La penultima ventura,* 96)

The seeds of such rhetoric, which reached its height in the Fiuman speeches (only to be stifled a few years later as Mussolini wisely if nervously subsidized D'Annunzio's lavish "exile" to the villa known as the Vittoriale), had thus already been sown in his "Canto augurale per la nazione eletta" in 1899.

CHAPTER FOUR

Biopolitics and Colonial Drive

Breached Walls and Wounded Bodies

It has often been remarked that Italian cinema was born "under the sign of the Risorgimento," itself a national "resurgence," or "rebirth" (Brunetta, *The History of Italian Cinema* 16). On September 20, 1905—the thirty-fifth anniversary of the Italian state's annexation of Rome—Filoteo Alberini's seven-frame film reenactment of the event, titled *La presa di Roma* (*The Taking of Rome*), was screened for an audience of thousands outdoors, adjacent to the very backdrop against which it was filmed.[1] A huge white screen was erected next to the famous "breach" of the Aurelian wall at Porta Pia, the widely mythologized point of entry for Italian troops in their so-called conquest of Rome.[2] In post-Unification Italy, rhetorical losses, ruptures, or "breaches," abound, and what they refer to most frequently are perceived challenges to Italy's territorial and corporeal integrity. Italy's constitutive fractures are perhaps the central preoccupation of Italian modernity, and rhetorics of loss (through the emigration of labor-power; racial decline or degeneration; corporeal and territorial mutilation) function as mechanisms of disavowal,

1 For a historical reconstruction of the film's production and release, as well as essays on its modern restoration, see Musumeci and Toffetti. See also Canosa, *La presa di Roma*. For film titles and citations, I have followed the notation style used by Bertellini in "Introduction: Early Italian Cinema" 235–239 and again in Bertellini, *Italy in Early American Cinema* 367.

2 In "L'entrata dell'esercito italiano in Roma," Edmondo De Amicis provides a first-hand journalistic account of the events of September 20, 1870, which already deploys a visual-cinematic style. His prose, which dwells for some time in the oneiric, glides through the central neighborhoods of Rome like a *carrellata,* or a tracking shot (*Roma capitale*).

Fig. 4.1 Alberini's *La presa di Roma* (1905).
The intertitle reads: "Attack! The Breach of Porta Pia"

paradoxically "mending" such fractures.[3] In the case of *La presa di Roma,* the territorial unification of the modern, secular Italian state (though still "incomplete," even after World War I) is affected *through* a breach; proclaiming its wholeness therefore requires a rupture (Fig. 4.1).

Four years later, in 1909, Alberini's second ode to the Risorgimento, *Il piccolo garibaldino* (*The Little Garibaldian Soldier*), narrates the ascent to martyrdom of a young patriot who dies in battle alongside popular Risorgimento hero Giuseppe Garibaldi. The closing scene is a dream-sequence of the dead young hero's mother, who solemnly accepts her son's allegiance to a new, spiritual mother, *l'Italia turrita,* as she consecrates his sacrifice by kissing his bloody flesh wound (Fig. 4.2). The "birth of the nation" that Italy's first films celebrate thus hinges on both a territorial rupture and a bodily one.[4]

3 Here I refer to Roberto Esposito's remarks, which serve as the epigraph to this book (Campbell, "Interview with Roberto Esposito" 49). See also the Introduction to the present study.

4 Here I refer of course to D.W. Griffith's groundbreaking film *The Birth of a Nation,* about which Michael Rogin has remarked: "American film was born in a racist epic" ("The Sword Became a Flashing Vision" 191). While the same cannot be said for these

Fig. 4.2 Alberini's *Il piccolo garibaldino* (1909)

Several rhetorical formulations that have emerged over the course of the past several chapters—from colonialism as a pedagogic, therapeutic, or life-affirming, rather than a violent and life-negating set of practices, to the rhetoric of territorial and corporeal loss and restoration—return in this chapter in an analysis of the encounter between biopolitics and early Italian cinema. A close reading of Giovanni Pastrone's 1914 epic silent film *Cabiria,* liberal Italy's first international blockbuster and a film that commemorated the Italian invasion of Tripolitania and Cyrenaica (Libya) in 1911, illustrates how the biopolitical constellation I have been tracing emerges in one of the most important technological innovations of the turn of the twentieth century, that magical encounter between industry and art known as cinema.[5] Between the close of the nineteenth century and the disappointing outcome of the Paris Peace Conference that "mutilated" Italy's victory in World War I and prompted D'Annunzio's occupation of Fiume, Prime Minister Giovanni

Italian "birth of a nation" short films, my goal in this chapter is to illustrate the links between their staging of territorial and bodily loss or rupture and turn-of-the-twentieth century Italian racial logics.

5 *Cabiria* is based loosely upon literary precedents: Emilio Salgari's *Cartagine in fiamme* (1908) and Gustave Flaubert's *Salammbô* (1862).

Giolitti had invested significant human and economic capital in an attempt to wrest the territories of Tripolitania and Cyrenaica from Ottoman rule and local resistance, beginning in October 1911.[6]

The Giolitti administration's decision to invade Libya was accompanied by a flurry of propaganda from Italy's political and literary frontline. Seven years before he occupied Fiume, D'Annunzio's verses "Canzoni della geste d'Oltremare" praised the "glory of the race" (*razza*) of Aeneas in its effort to conquer Libya and "widen [Italy's] skies."[7] In the *Canzoni*, which Lucia Re argues are a turning point in Italian racial discourse because they reflect the passage from a nineteenth-century, positivist discourse on race as divisive and polarizing to a unified, "modern," twentieth-century "colonial and racial national identity," D'Annunzio, "traces an imaginary, poetic hyper-map across the Mediterranean that connects disparate events in time and space in order to highlight '*la gloria della razza*'" ("Italians and the Invention of Race" 28). At the end of November 1911, poet Giovanni Pascoli delivered one of his last public addresses in praise of "proletarian" colonization to a crowd in

6 At the beginning of that month, 35,000 Italian troops landed in Tripoli furnished with neither maps nor interpreters on the hunch that the scarce number of Ottoman troops stationed there would pose no threat. Instead, the invasion would mark the beginning of a brutal twenty-year campaign for Italian control, punctuated by the first use of aircraft in a military offensive, the deployment of mustard gas, the violent expropriation of agricultural land from local inhabitants of Cyrenaica and their internment in concentration camps in the Syrtic desert, and culminating in the capture of septuagenarian and Cyrenaican resistance leader 'Umar al-Mukhtar and his public execution by hanging in 1931. In 1934, Tripolitania and Cyrenaica, two regions divided by the Syrtic desert, which had historically been distinguished by two distinct spheres of influence: the Maghreb for the former and Egypt for the latter, were united (along with the desert of Fazzan) to form a single colony, which was named Libya—a term classical texts used to designate African regions west of Egypt. Rochat quips that this was the only act of Italian colonialism with which Libyan dictator Muammar el-Qaddafi has never taken issue. Historians Ali A. Ahmida (*The Making of Modern Italy*), Angelo Del Boca (*Gli italiani in Africa Orientale*), Nicola Labanca ("The Embarrassment of Libya"), and Giorgio Rochat (*Le guerre italiane in Libia e in Etiopia*) have exposed and denounced many of the details about this episode in Italian history, which the Italian state had for many years obscured. For a reading of the Libyan "Arab Spring" in light of Italian colonial history, see Fuller, "Libyan Genocide 2.0." Fuller suggests that just one of the ironies of today's Italo-Libyan relations is that: "Qadhafi has already replaced Italians in Libyan memory as the perpetrator of genocide."

7 "La canzone del sangue," 29. The *Canzoni* were published in the *Corriere della Sera* between October 1911 and January 1912, and are now contained in *Merope*, book four of *Laudi del cielo, del mare, della terra, e degli eroi*.

Barga; the speech also appeared in the Roman daily *La Tribuna* (*La grande proletaria si è mossa*). Futurist F.T. Marinetti sent dispatches from the Libyan front to the Parisian newspaper *L'Intransigeant,* which he later published in Italian as *La Battaglia di Tripoli* (1912).[8] Enrico Corradini, founder of the Italian Nationalist Association (ANI), published three volumes, *L'Ora di Tripoli, La conquista di Tripoli,* and *Sopra le vie del nuovo impero,* in which he shared his vision that one day, "with strength, activity, and the conquest of African treasures, a whole new population with Italian blood" would occupy Mediterranean Africa.[9] Earlier in 1911, Corradini, along with Alfredo Rocco and Luigi Federzoni (later appointed Minister of the Colonies) had anticipated the Italian invasion, issuing the first edition of their newspaper *The National Idea* on March 1, the twenty-fifth anniversary of the Italian army's historic defeat at Adwa (Labanca, *Oltremare* 112).[10] Colonial defeats—at Dogali in 1887 and at Adwa in 1896—might be considered additional coordinates in the constellation of rhetorical loss that I have been tracing. These defeats had carried significant rhetorical weight since the earliest utterances of pro-colonial propaganda, exemplified in the late nineteenth-century works of Alfredo Oriani.[11] Overwhelmingly, loss, rather than victory, served to mobilize public opinion in favor of colonial war.

D'Annunzio's speech to aviators at Centocelle in 1919 continued to rehearse this narrative, in which the Italian defeat at Adwa would be avenged by Italian colonial victory elsewhere. For D'Annunzio and other colonial enthusiasts, the precise location of this elsewhere was less important than the tenor of the promise that there was still land left to conquer on Italy's "fourth shore" in the Mediterranean.[12] As he would make clear on the eve of his occupation of "unredeemed" Fiume with his image of an airway fusing the distant and

8 The preface to the collection includes a manifesto published at the beginning of the conflict with an exhortation to the "poets, painters, sculptors and musicians of Italy" to put aside "verses, paintbrushes, chisels and orchestras" and turn to admire instead the "formidable symphonies of shrapnel and the wild sculptures that our inspired artillery forges within enemy masses" (Marinetti, *La battaglia di Tripoli*).

9 Corradini, *La conquista di Tripoli* 56 (cited in *Tomasello, L'Africa tra mito e realtà* 52). See also Corradini, *L'Ora di Tripoli*; *Sopra le vie del nuovo impero*.

10 For a rich account of the contours and breadth of the colonial imaginary in Italy from the 1880s through World War II, see L. Ricci.

11 See Chapter One of the present study.

12 Labanca notes that more important than Rome's control of Libya was avoiding the risk that Libya fall into the hands of London or Paris, which would have meant Italy's geopolitical exclusion from the southern Mediterranean (Labanca, *Oltremare* 109).

disparate lands of Libya, Eritrea, and Somalia ("Make it so that between Tripoli and Murzuq, between Tripoli and Tumu, the yellow road transforms itself into a sky-blue road, that the oasis at Kufra sees the tri-colored wings arrive through the air like flags hoisted to honor the new *Patria*. Connect Tripolitania to Eritrea, Cyrenaica to Somalia," *La penultima ventura*, 98), and as colonial historians have maintained, internal differences and geographical distance between the colonies were irrelevant to a colonial politics based less upon access to resources specific to desired lands and more upon an abstract notion of conquest, or as colonial historian Nicola Labanca has put it, "the search for international prestige, [or] the politics of might [*potenza*]" (*Oltremare* 109).[13] By the last decade of the nineteenth century, Italy's colonies in the Horn of Africa had proved incapable of attracting the large numbers of Italian emigrants Franchetti and others had hoped. Increasingly, beginning with the Italian military's initial incursion upon Libyan shores and continuing through the 1930s, when the fascist government sent Italians to settle some thirty villages there, hopes for the agricultural self-sufficiency of poor Italian peasants would be directed at the only portion of the North African coast that had not yet been occupied by France or Great Britain.[14] The vagueness of Italy's geographic coordinates of conquest and their eventual establishment through deduction (the territorial table scraps of the Berlin conference) lent a specificity to the Italian colonization of Africa, which Karen Pinkus has characterized in the following way: "Italy's peculiar indolence in solidifying the borders that would serve to frame the black but that also would expand and turn slowly green (the color of agricultural reclamation, as well as unified Italy), is key" (29). Pinkus' claim invites us to think about the chromatics of colonial conquest in this period not only in black and white, but also in the green of a promised land of abundant crop cultivation. What Italians found instead, with the exception of the verdant areas along the Libyan coastline, turned out to be "an immense sandbox" characterized instead by the beige of harsh and uncultivable desert sands.[15]

13 For a similar claim about the Italian government's superficial knowledge of the regions and people they set out to conquer, see Rochat, "Il colonialismo italiano"; *Le guerre italiane in Libia e in Etiopia* 24.

14 Fuller points out that, although "the preferred image of the colonies portrayed a promised land full of Italian farmers," the only sectors that grew significantly over the course of the Italian colonial enterprise were the military and government bureaucracy (*Moderns Abroad* 38).

15 Fuller notes that the few voices of anti-colonial dissent rehearsed Prime Minister Francesco Nitti's disparaging remark that Libya was merely "an immense sandbox"

The widely circulated literary-propagandistic works in praise of Libyan conquest by Corradini, D'Annunzio, Pascoli, Marinetti, and others signaled a critical juncture on the road toward the aestheticization of politics that D'Annunzio would rehearse at Fiume and—as Walter Benjamin famously recognized—fascism would attempt to perfect, through, among other things, its efforts to exercise ultimate control over the content and distribution of popular culture in general and the film industry in particular.[16] Poets and literary propagandists found much-needed inspiration in Libyan conquest (united Italy's first and longest national war[17]) as it served variously as an opportunity to confirm the aesthetic value of battle (Marinetti), to revive ancient Italian primacy through heroism and sacrifice (D'Annunzio), to deliver Italy's huddled masses from being "swallowed up" by other nations (Pascoli), and to enable the "proletarian nation" to conduct its "exuberant genetic strength [*potenza*]" toward the elimination of inferior civilizations (Corradini).[18] The invasion of Libya in 1911 also marked a turning point

(*Moderns Abroad* 47). Augusto Genina's Mussolini Cup-winning 1936 film about the Italian occupation of Tripolitania, *Lo squadrone bianco* [*The White Squadron*, Roma Film] relies heavily upon the chromatic binarism of black and white, as its title suggests. The blinding sun and sand of the Libyan desert also play a critical role in the conversion of the decadent bourgeois young man into a proper colonial hero Mario (Antonio Centa) and the demise of his predecessor, Capitan Santelia (Fosco Giachetti). See: Bertellini; Boggio.

16 For a compelling reading of film history that disrupts the by now conventional, Benjaminian link between film and modernity by highlighting instead cinema's reliance on "pictorial statements"—particularly those that narrativize race—which predate the typical hallmarks of modernity such as industrialization, mechanical reproducibility, and mass culture, see "A Mirror with a Memory," in Bertellini, *Italy in Early American Cinema* 276–291. Volumes have been written on Italian film production under fascism. Seminal works in Italian include: Argentieri; Brunetta, *Cinema italiano tra le due guerre*; Redi. In English, see: Ben-Ghiat, *Fascist Modernities*; Landy; Reich and Garofalo; S. Ricci.

17 Giorgio Rochat describes how the Italian military distinguishes between "national" and "colonial" wars. "National" wars require large government expenditure, the direct involvement of the Italian Army, and widespread public support ensured through propaganda campaigns. Examples are the 1911–1912 war in Libya and the 1935–1936 war in Ethiopia. "Colonial" wars, on the other hand, were administered by the Ministry of Colonies (later the Ministry of Italian Africa), involved a reduced number of Italian forces, and enjoyed little to no public support (*Le guerre italiane in Libia e in Etiopia* 23).

18 These distilled positions refer to D'Annunzio, Pascoli, and Marinetti's Libyan works, cited in the text and notes above. Pascoli's formulation was that Italian emigrant laborers "were lost in the whirlpool [*gorghi*] of other nationalities" (*La grande proletaria si è mossa* 9). For Corradini, see *Sopra le vie del nuovo impero* 47–48 (cited in Tomasello, *L'Africa tra mito e realtà* 53).

in Italy's cinematic conquest of Africa.[19] The invasion was immediately relayed to Italian audiences in New York through the documentary film *Guerra in Tripolitania* (*Italian-Turkish War,* Cines, 1911) (Bertellini, *Italy in Early American Cinema* 12). The war gave what preeminent historian of Italian cinema Gian Piero Brunetta calls "a decisive push to the confluence of cinematic production, nationalist ideology, and the ambitions of a small country that had been unified and liberated only a few decades earlier" (*History of Italian Cinema* 34). Film pioneer Luca Comerio (1870–1940) accompanied Italian troops as they landed at Tripoli, marking a historic union between war and cinema that would be further solidified with his appointment as the Italian army's official filmmaker during World War I, and some ten years later with Mussolini's notorious maxim that "cinema is the [regime's] strongest weapon."[20] Anticipating Marinetti's "Manifesto of Futurist Cinema" by some five years, Comerio allegedly strapped himself to an airplane in order to film from the very Libyan sky that Marinetti would later visit (Moliterno 90). The influence was mutual: as Brunetta points out, Marinetti's earlier literary works, particularly his aeronautical and anticlerical poem *Il monoplano del Papa* (*The Pope's Monoplane*) of 1910, already deployed a cinematic gaze that would share affinities with that of Comerio's films (*The History of Italian Cinema* 34). In addition to Marinetti and Comerio, notable early filmmakers such as Giovanni Vitrotti and Elvira Notari found Libya an apt stage for ethnographic films, war reportage, and popular drama. Notari's *L'eroismo d'un aviatore a Tripoli* (*Heroism of an Aviator in Tripoli,* Dora Film, 1912) decried the Libyan war's obliteration of enemy life and domestic amorous passion: "In Libya the war showed the terrible potential of aviation: the rain of bombs exploding confused and destroyed the enemy. Strange spectacle! Powerful means of destruction," announced the film's synopsis, referring to the fact that Italian aircraft

19 For a historical overview of narrative films produced in Italy and set and/or filmed in Africa, including a useful filmography, see Brunetta and Gili. For a more comprehensive filmography of Italian films on Africa, as well as a discussion (in English) of Italian films and newsreels on Africa during and after fascism (1930–1960), see Baratieri, whose useful historical study, despite several regrettable editorial errors, fills a void in scholarship on Italian film about Africa. Indeed, since around 2010 a veritable flurry of film scholarship has emerged that seeks to redress longstanding lacunae in the (post) colonial chapter(s) of Italian film history. In addition to the pioneering works by Ruth Ben-Ghiat and Giorgio Bertellini, see: Franceschi; Greene; O'Healy; Trento; Duncan, "Italy's Postcolonial Cinema."

20 For more on Comerio, see: Comerio et al.; Dagrada, et al.

had bombed Tripoli in November 1911 (an event which made history as the first use of aircraft in a military offensive).[21] To paraphrase Paul Virilio (writing about World War I), during the Libyan occupation, war, aviation, and cinema formed "a single moment," influencing visual perception for soldiers and directors alike (4). I will return to a fuller discussion of this below, but for now it will suffice to mention that Pastrone's *Cabiria* stands apart from these other texts insofar as it takes up Libya as neither a site for visual war reportage, nor as a racy backdrop for a contemporary drama or comedy. Instead, Pastrone—following literary-propagandistic precedents set by Corradini, Pascoli, and D'Annunzio, and in line with the early Italian film industry's fascination with antiquity—conjures up a mythical past in order to justify the present by staging an analogy between ancient Rome's battle for Carthage and contemporary Rome's attempts to colonize Libya.[22]

21 *La vita cinematografica*, Turin, n. 20, 1912, p. 35 (cited in Bruno 199). Bruno notes the scholarly neglect of the silent film era in Italy (with the exception of grand literary and historic epics such as *Cabiria*), which led to the obscurity of regional filmmakers like Notari, despite her prolific filmography and international success (14). Historical studies of the Italian silent era include: Bernardini; Prolo; Canosa, *Cinema muto italiano*. For a technical and historical overview of the silent era, including the transition to color and the preservation of the film stock, as well as a filmography and a guide to archival silent film research, see *Cherchi Usai, Silent Cinema*. Useful anthologies on the American and European silent traditions include: Abel, *Silent Film*; Grieveson and Kramer, *The Silent Cinema Reader*.

22 Though Marinetti was one of the loudest voices in this pro-colonial chorus, his disdain for "passatismo" meant that his Libyan dream did not rely upon conjuring up a mythic return to ancient Rome. In October 1911, he published his "Second Futurist Political Manifesto," in which he celebrated the Libyan war by declaring that, among other things, "The tiresome memory of the greatness of Ancient Rome must be eradicated by an Italian greatness that is a hundred times more impressive" (*Critical Writings* 73–74). The piece first appeared in *Marinetti, Guerra sola igiene del mondo*. It appears as "Manifesto a Tripoli italiana" in Marinetti, *Futurismo e fascismo*. In an interview with Sicilian periodical *L'avvenire* in February 1915, Marinetti explained his distance from Corradini and the Nationalists for wanting to "consolidate the nation around tradition, a mania for monuments, and veneration for ancient ruins. [...] I am with them, however, as brothers-in-arms, when it's a case of spitting in the face or kicking the living daylights out of all those pacifists and socialists who would like to sully, humiliate, and degrade this great word *Italy*. I want to make the point, however, that the Italy of tomorrow must be, and will be, infinitely greater than the archaeological and cultural one the nationalists are hell-bent on cobbling together, restoring, and setting up on pedestals" (*Critical Writings* 240).

Circuits of Loss and Return

The critical history of Pastrone's pioneering film has been shaped by an overwhelming concern with the film's technological achievements and difficulties. *Cabiria* is remembered by film scholars for its innovative use of the tracking shot, and more specifically for the use of the dolly mounted on a sinusoidal track to emphasize the three dimensionality of cinematic space (Cherchi Usai, "*Cabiria,* an Incomplete Masterpiece" 161). Like the set's dimensions, the film's set designs and pyrotechnic effects have been praised for their monumentality. D.W. Griffith is said to have raced to San Francisco to attend a screening of Pastrone's masterpiece sometime after completing *The Birth of a Nation* (1915). Griffith was allegedly so humbled by Pastrone's cinematic achievement that his own appeared "primeval" in comparison. He would go on to shoot *Intolerance* (1916) determined to outdo his Italian contemporary.

Equally intriguing for scholars of *Cabiria* has been the elusiveness and fragmentation of the filmic material itself. At least until 1988, there were multiple versions in circulation, most based upon a truncated re-issue of the film in 1931. The original 1914 version—released at over 3 hours—was rumored to have been lost in a fire. The quest that ensued for the lost film of *Cabiria* thus parallels the film's narrative itself, which recounts the battle between Rome and Carthage during the Second Punic War by following the somewhat accidental capture of the title character, a young Roman girl named Cabiria, her subsequent sale at a Carthaginian market, her near sacrifice to the fire god Moloch, and her eventual rescue and return to Rome by the darkened slave Maciste. In the late 1980s, film scholar Paolo Cherchi Usai dramatized this technological and narrative overlap when he called for a restoration of the original 1914 version by playfully suggesting, "We know that parts of the lost [film] *Cabiria* are still retrievable, but it would be premature to say that the little Carthaginian slave has yet been saved" ("*Cabiria,* an Incomplete Masterpiece" 165).[23]

Cabiria is a film that posits a territorial "loss"—formerly Roman territories on Italy's "fourth shore"—as remedied by a corporeal return. In addition to staging a parallel between territorial and bodily restitution, Pastrone's rhetoric of loss and return, like the nebulous imaginative geographies of Italian colonialism I mentioned above, requires a temporal and geographic conflation: contemporary Tripolitania and Cyrenaica stand in for ancient

23 Cherchi Usai's call was heeded, as *Cabiria* was restored once in 1995 and again in 2006 (de Oliveira).

Carthage (in modern Tunisia, where the film's desert scenes were in fact filmed). To reference Carthage at the onset of the Libyan campaign was to refer to the far less ancient "loss" of Tunisia to France roughly thirty years prior. I've pointed out that two colonial defeats, those of Dogali and Adwa, played a large role in bringing public opinion around in favor of colonial conquest. To these losses one might add several others—both real and imagined—that lent shape to colonial rhetoric in the period between Unification and the close of World War I. The loss of Italian bodies to the Americas that early colonialist Leopoldo Franchetti used to justify "demographic" colonialism in Eritrea was figured as both economic and biological; lest they were redirected to colonial lands, Italian emigrants had no choice but to "revitalize" other nationalities, financially and racially.[24] France's colonization of Tunisia in 1881 brought yet another perceived loss to the attention of Italians who were worried about their place in the imperial world order. Beginning in the 1820s and throughout the wars of Italian independence and unification, significant numbers of Italians representing all sectors of the economy settled in Tunisia, constituting the largest European community there.[25] By the time France declared Tunisia a protectorate in 1881, there were tens of thousands of Italians (primarily Sicilians) working and living there who, despite the fact that they heavily outnumbered the French, stood to become their subjects (Brondino 16–21). In the words of Franchetti, writing after the Italian invasion of Libya, "In the remaining regions around [the Mediterranean Sea], occupied by other civilized nations, Italian activity is taking place, but our compatriots are destined to be absorbed sooner or later by the dominating nationality" ("Prefazione" iv). Franchetti's language once again betrays a concern with what might be called the liquidity of political subjection, which in turn recalls a biological, if not reproductive, model according to which active Italian laborers risked being "absorbed" by "the dominating nationality"; to what does this absorption refer if not a kind of miscegenation between Italians and those "other civilized nations" who stood to dominate them?

24 Likewise, as I argued in the preceding chapter, the melancholic hero of D'Annunzio's novel *Trionfo della morte* (1894) Giorgio Aurispa figures his father's financial ruin as a bloodletting, producing yet another figure for economic and biological loss, which is tied to Aurispa's preoccupation with racial decline.

25 Most Italians settled in Tunis and other coastal centers such as La Goulette (or "Goletta"), Bizerte, and Sfax and worked as merchants, industrialists, doctors, lawyers, local administrators, and military personnel, as well as fishermen, farmers, and manual laborers. See: Melfa; Choate, *Emigrant Nation* 83–86; Marilotti; Pendola; Rainero, *Les Italiens dans la Tunisie contemporaine*; Rainero, *L'Italia e il Nordafrica contemporaneo*.

If Eritrea was figured in reproductive (and doubly gendered) language as Italy's "first-born colony" (*colonia primogenita*), colonial proponents overwhelmingly described Libya as a site of return. D'Annunzio's verses in praise of Libyan occupation refer to the return of a "fertile race" (*stirpe ferace*) to Italy's "fourth shore," upon which it would "once again deepen its ancient trace" (D'Annunzio, "La canzone d'Oltremare" in *Merope* 650). Ancient Roman architecture at sites such as Leptis Magna were presented as proof that, in the memorable words of Giovanni Pascoli: "ci fummo già" (we were already there). "We left traces that neither Berbers, Bedouins nor Turks could erase," Pascoli claims. The "return" to Libya would accomplish what the battles of the Risorgimento had not: "Now fifty years ago, Italy was made. On the sacred fiftieth anniversary, you have also proved [...] that Italians were also made," concludes Pascoli (*La grande proletaria si è mossa* 18; 25). This rhetoric of return was mobilized, of course, to justify even more loss. Predictably, thousands of war dead and wounded (on both sides of the conflict) were vital to accomplishing Italy's triumphant "return" to Mediterranean Africa. If colonial rhetoricians cashed in on the affective value of loss and return, it was because it appealed to a preexisting set of societal anxieties about the circulation of Italian bodies and the land they either inhabited or sought to. As Lucia Re describes:

> Some of the most jingoist accounts of the Libyan war, built on the double myth of 'the return,' were in fact published in Latin America to appeal to the emigrant community. The colonial war was represented as a return of Italians to the land beneficially colonized by their Roman forefathers, and the conquest of Libya by the motherland would provide the opportunity for the victims of the Italian diaspora finally to return home to 'their sea,' the Mediterranean. ("Italians and the Invention of Race" 34)

The territorial breach that brought about the "birth of the [Italian] nation" in Alberini's 1905 film, coupled with the bodily rupture of the young Garibaldian soldier that concluded Alberini's second film likewise registered such national anxieties about bodily and territorial fragmentation. Alberini's short films dramatized how the acquisition of Rome that "completed" the territorial unification of Italy required the breach of the wall at Porta Pia; likewise, the corporeal unification of Italy (which was urgent precisely because of so much bodily loss—to emigration and battle—that preceded it) required a flesh wound. The phantasmagoric restoration of both bodies and lands, national and colonial, were thus inscribed in an economy of loss.

Colonial Drive

If anxiety shaped the Italian colonial imaginary at the turn of the century, so did desire. Following Frantz Fanon, Homi Bhabha, and Achille Mbembe, who in distinct ways have demonstrated what is to be gained from analyzing colonialism in psychoanalytic terms, cultural studies of colonialism have dedicated volumes over the past several decades to contextualizing and theorizing desire in the colonial context. Numerous studies of French, British, Spanish, and to a lesser extent German colonialism take up real or imagined sexual and affective relationships between colonizer and colonized, between desiring subjects and their objects of desire.[26] Many important works inspired by these readings of the admittedly more robust colonial traditions of other nation-states have been directed at the Italian case, unpacking how the colonial encounter is both structured by and structures sexual and affective bonds, dreams, fantasies, and so on.[27] Such work has been fundamental in accounting for how—far from being carried out unilaterally, with sheer force by gun-wielding white men—colonialism required an array of often intimate alliances between men and women, subjects and objects, whites and non-whites. While I have no reservations about the advantages of addressing the colonial encounter through the analytic of Freudian desire in the Italian case (one need only think of the popular tunes "Faccetta nera" and "Tripoli, bel suol d'amore" to get a sense of how Italian colonialism blatantly conflated sexual and territorial conquest) it may be more apt to think of the Italian preoccupation with loss that I have been discussing throughout this book's chapters through Lacan's concept of drive, as he distinguishes it from desire. While, for Lacan, desire is a dialectically (and ideologically) constituted repetitive loop that continues to produce only more of itself ("a desire to desire," as Jodi Dean paraphrases it), the subject of drive attains enjoyment (*jouissance*) through repeated failure. Slavoj Žižek explains:

26 A few recent examples include: Holden and Ruppel; Yegenoglu; Young, *Colonial Desire*; Zantop.
27 Bringing an analysis of Italian cinema of the 1930s to bear on the colonial context of desire, Cecilia Boggio has gone so far as to claim, "The truth is that Italians were controlling Ethiopia only with and through cinematic fantasies" (279). While historians might take issue with such an understanding of colonial fantasy—citing, among other things, the deployment of mustard gas, which caused the violent deaths and subjection of local populations—cinema was one of the perceived "weapons" of the regime in the colonies, aimed at both local populations and colonials alike. See Ben-Ghiat, "The Italian Colonial Cinema," *Italian Fascism's Empire Cinema*.

> [A]lthough in both cases, the link between object and loss is crucial, in the case of the *objet a* as the object of *desire,* we have an object which was originally lost [...], whereas in the case of the *objet a* as the object of drive, the "object" is directly the loss itself—in the shift from desire to drive, we pass from the *lost object* to *loss itself as an object.* That is to say, the weird movement called "drive" is not driven by the "impossible" quest for the lost object; it is *a push to directly enact the "loss"—the gap, cut, distance—itself.* (Emphasis in original. *In Defense of Lost Causes* 328)[28]

As texts by Franchetti, Mantegazza, and D'Annunzio have revealed, corporeal loss, conceived of as displacement, dismemberment, decline, and/or death, was a profoundly Italian cultural preoccupation in the historical moment under consideration, particularly given the fact that, in the case of emigration, the bodies dislodged from their roots in Italian soil had, more often than not, never been "Italian" to begin with (they left the peninsula as Genoans, Lombards, Venetians, Neapolitans, and so on). The outpouring of literary depictions of such emphatically national loss—from at least as early as Ugo Foscolo (*Le ultime lettere di Jacopo Ortis* and *Dei sepolcri*), and Giacomo Leopardi (the *Canti* of 1818) to Edmondo De Amicis ("Dagli Appennini alle Ande" and *Sull'oceano*), Giovanni Pascoli ("Italy") and Gabriele D'Annunzio ("Agli italiani degli Stati Uniti")—served only to tighten the knot of signification, ensuring the fetishization of loss itself.[29] Colonialism was presented as therapeutic, a paradoxical way of healing fragmentation through sacrifice and bodily mutilation. In Pascoli's words:

> Terra, mare e cielo, alpi e pianura, penisola e isole, settentrione e mezzogiorno, vi sono perfettamente fusi. Il roseo e grave alpino combatte vicino al bruno e snello siciliano, l'alto granatiere Lombardo s'affratella col piccolo e adusto fuciliere sardo. [...] Scorrete le liste dei morti gloriosi, dei feriti felici della loro luminosa ferita: voi avrete agio di ricordare e

28 Lacan's *jouissance* is critical to Žižek's notion of politics. Jodi Dean writes that, with *jouissance,* "he is calling our attention to the way that we all, in contemporary consumer-driven entertainment society, enjoy popular culture and the way this enjoyment binds us into the ideological formation that supports global capital" (*Žižek's Politics* xvi).

29 Lacan's *objet petit a,* on which Žižek's reading of drive hinges, is similar to Freud's fetish. Whereas Lacan's *objet petit a* stands in for a lost object (the mother's breast), Freud's fetish stands in for an object that was never there in the first place (the phallic mother).

> ripassare la geografia di questa che appunto era, tempo fa, una espressione geografica. (*La grande proletaria si è mossa* 15)
>
> [Land, sea and sky, mountains and plains, peninsulas and islands, north and south, they are perfectly fused. The rosy and grave Alpine fights near the thin, brown Sicilian; the tall Lombard grenadier swears brotherhood to the small, sallow Sardinian rifleman. [...] Scan the lists of the glorious dead, of the wounded, rejoicing in their luminous wounds; you will immediately be able to review the entire geography of Italy, Italy that a short time ago existed only as a "geographical expression." (Baranello 11)]

Territories, like the racialized bodies (the ruddy northerner, the swarthy southerner) that inhabit them, once divided, are miraculously soldered together through colonial war. In appealing to colonialism as a chance to mend these biopolitical fractures, a kind of colonial drive is activated, in which colonial defeat as a justification for colonial war ensures loss follows ever more loss. In such a rhetorical context, colonial battle wounds become a source not of pain, but of patriotic pleasure; they are "luminous." Wounds, defeat, loss—these are the keywords of early Italian colonial discourse that entrap it in the circuit of drive. For Jodi Dean, drive is the structure of biopolitics itself, which accounts for how, in its attempt to protect life, it winds up in "monstrous reversals, as biopolitics turns into an intensified politics of death."[30] Indeed, in post-Unification Italy, Dean's biopolitical drive played itself out in a variety of ways. For Franchetti, the "vital forces of the nation" are to nourish Italian soldiers as they defend colonial borders in Eritrea; Mantegazza's regulatory future of global reproductive health necessitates the incineration of infants who are deemed biologically unfit; and D'Annunzio's triumphant regeneration of the race at Fiume demands Italian blood be spilled and bodies mutilated in battle.

30 For Dean, atomic power and racism exemplify the loop of biopolitical drive, "the turning round and reversing, the movement outward and back" that results in capture. On atomic power as an "externalized biopolitics," she writes: "[I]tself the manifestation and result of the imperative to secure a population completely, atomic power can, if deployed, destroy the sovereign power to maintain life." On racism as an "internalized biopolitics," she writes: "Racism is bound up with the technologies to put biopower to work, with a State that uses race to justify its sovereign power of making life and letting live and so mobilizes death in order to strengthen the race" ("Drive as the Structure of Biopolitics" 6).

A Peculiar Blackness

> [T]he historical index of the images not only says that they belong to a particular time; it says, above all, that they attain to legibility only at a particular time. [...] Every present day is determined by the images that are synchronic with it: each 'now' is the now of a particular recognizability.
>
> Walter Benjamin, *The Arcades Project* (462–463)

Biopolitical discourse in post-Unification Italy involves primarily physiological expressions of race; in the cultural production explored in this book, the discursive project of "making Italians" has been comprised most often of elements such as blood, organs, and vital fitness. Likewise, whiteness has either been presumed "neutral"—what Ross Chambers has called "the unexamined"—or it has been feverishly insisted upon as an indicator of physiological decline (here I refer to D'Annunzio of the preceding chapter, for whose women this constitutes an aesthetic value) or in contrast to blackness, as Chapter One demonstrated with Enrico Persano's description of Italians as "white men" inhabiting new colonial climates, and in Chapter Two with Mantegazza's assignment of Italians to a higher level of psychic civilization than "negroes."[31] Often, more important than skin color has been a unique preoccupation with which bodies are reproducing, laboring, or sacrificing themselves in battle, and where—be it in the Americas, Italy's southern regions and islands, the new colony in the Horn of Africa, the mutilated territories of the Adriatic, the lost ones of the Mediterranean, or the science-fictive capital of the United Planetary States in the year 3000, Andropolis. In Pastrone's *Cabiria*, a peculiar, simulated blackness that is emphatically

31 Chambers' essay, which, significantly, begins with an anecdote about his "discovery" of the Sicilian provenance of a flight-mate ("a Mediterranean" with "black hair, olive complexion, dark eyes") from Sydney to Rome, argues that whiteness' taken-for-grantedness can be summed up in the word "in(di)visibility." While in a variety of contexts (Chambers mentions Australia, South Africa, and the United States) nonwhites are homogenized through plurality, whites enjoy a quality of "undividedness" and "aparadigmacity," or neutral wholeness that enables an individualized—rather than plural and homogenized—identity.

not the racist blackface of the late-nineteenth century American and British traditions instead comes into view.[32]

Italy's incursion into Tripolitania and Cyrenaica brought epidermal blackness (as well as Arab-ness) into sharper focus for an Italian public who, up until that point, had maintained a comfortably distant relationship to exoticized Africans gleaned primarily through the armchair adventure novels of Emilio Salgari and accounts of nineteenth-century travel writers, geographers, ethnographers, and positivist social scientists of a Lombrosian stripe. As Pinkus explains, "[B]lackness did not become a repository for any *particular* fantasies or cultural positions in Italy until the Libyan conquest of 1911" (29). Lucia Re has also argued that the Libyan campaign mended the fragmentary nature of Risorgimento-era Italian racial discourse, providing an 'other' upon which to project a unitary Italian race ("Italians and the Invention of Race" 3). *Cabiria*'s Maciste was played by Bartolomeo Pagano, a Genoan stevedore-turned-actor, whose skin was conspicuously darkened for his role as the noble slave (Fig. 4.3).[33]

32 In the follow-up to *Cabiria*, directed by Vincenzo Dénizot and Romano Luigi Borgnetto and titled simply *Maciste* (Itala Film, 1915), Pagano as Maciste theatrically disguises himself in minstrel-style blackface in order to rescue yet another damsel in distress. Jacqueline Reich reads this scene as confirming Maciste's whiteness by way of highlighting the comic performativity of his blackness in both this film and, by association, in *Cabiria* (249–251). Scholarship on blackface and minstrelsy in the U.S. and Britain is vast. See: Lhamon; Eric Lott, *Love and Theft*; Pickering; Rogin, *Blackface, White Noise*; Strausbaugh.

33 In a letter to D'Annunzio, who added rhetorical flourish to Pastrone's intertitles and notoriety to the film's publicity posters, Pastrone reveals a conscious decision to make the Maciste character a "mulatto," though he provides no indication of his motivation for this choice. In her article, Antonia Lant reads Pastrone's film as an illustration of two divergent approaches to the theoretical question (posed by Vachel Lindsay, Sergei Eizenstein, and Siegfried Kracauer) as to whether cinema was the twentieth-century reincarnation of the hieroglyph. Lant identifies two strands of thinking on the question, one originating in the U.S. and the other in Europe. The approach to the question that emerged in the U.S. was the capacity of cinema to represent the racial stratification, or perhaps more accurately polarization, that structured American life at the turn of the century. European filmmakers tended to address cinema-as-hieroglyph via their approach to three-dimensional cinematic space. Lant reminds us that cinema emerged not only at the height of western European imperialism, but also in the midst of a larger debate about spatiality in art in general, and in particular in the choice of avant-garde artists (Lant cites Paul Gauguin, Henri Matisse, and the cubists) to set aside, among other things, the constraints of Renaissance perspective. Orientalist visual topoi were thus dear to early European filmmakers not only for their capacity to reflect imperialist

Fig. 4.3 Maciste and Fulvio Axilla in Pastrone's *Cabiria* (1914)

The spectacle of Maciste's body—exposed, dark flesh and glistening muscles—was one of the primary reasons for the film's enormous success. Pastrone and others capitalized on Maciste's popularity, as Pagano went on through to the late 1920s to star in over twenty subsequent "Maciste" films, which would be the precursors to the wildly popular Italian "sword and sandal" or peplum films of the 1950s and 60s (Farassino, "Maciste e il paradigma divistico"; Schenk 160). If the peplum genre exploited classical antiquity in producing cinematic models of male masculinity, in Italy, this genre grew out of a circus tradition that by the mid-nineteenth-century had already begun to stage heroic gladiators alongside elephants and tigers in performing Biblical and classical texts. The development of a parallel, if not overlapping, "strongman" genre characterized by feats of athleticism and acrobatics can also be traced to Maciste's role in *Cabiria* (as well as Bruto Castellani's Ursus from Enrico Guazzoni's 1913 colossal *Quo Vadis?*) (Farassino, "Anatomia del cinema muscolare" 30). While scholars have acknowledged *Cabiria*'s Maciste as the "father" of subsequent muscle-bound heroes such as Rambo and Conan, they have tended to focus their attention on the construction of these later heroes as necessarily white (Rushing). One

ambitions, but also for their ability to add a monumental and ornate depth to the cinematic canvas.

feature of Maciste's appearance that has recently gained attention by film scholars such as Giorgio Bertellini ("Colonial Autism"), Jacqueline Reich ("Slave to Fashion"), and Shelleen Greene (*Equivocal Subjects* 14–49) is the sudden "whitening" of Maciste—after *Cabiria,* Pagano's Maciste shed his dark skin, "evolving" from the darkened slave in *Cabiria* to an unequivocally white idealized Italian.[34]

Bearing in mind both Maciste's rather abrupt whitening and the rhetorical economy of territorial and corporeal loss that pervaded this historical moment in Italy, the artificiality of Maciste's darkness might be read within the film's visual and ideological registers of colonial war. *Cabiria* should be situated within a biopolitical context that not only anticipates the fascist aesthetics and politics of empire, but also includes the constellation of turn-of-the-twentieth-century concerns about the biological and territorial integrity of the Italian peninsula that I have been exploring over the past several chapters. *Cabiria* is a film about the "lost," set in the formerly Roman lands of northern Africa, whose narrative hinges on the loss and restoration of a proper Italian body—a dazzlingly white young blond-haired girl.[35] Robert Rushing has ventriloquized the film's underlying logic as, "Yes, perhaps colonialist exploitation is brutal, but, after all, we had to save that little girl" (173; 89 n. 28). Colonial war in the film is thus a justification for returning an Italian body to its "proper" home; this bodily loss and return parallels the loss and restoration of a colonial territory to Rome that was purportedly to be carried out as a result of Libyan conquest.

In addition to the critical attention it has received for its innovative use of the tracking shot and the monumentality of its sets, *Cabiria* is often described

34 Jacqueline Reich argues that Maciste's whitening after *Cabiria* runs parallel to his passage from a bare-chested strongman to a modern, bourgeois, suited (and eventually uniformed) hero in subsequent Maciste films. For Reich, Maciste's whitening resulted from a racialized notion of Italian nationalism that privileged the white male body, as well as the clothing that never fully masked its musculature, as the loci of virile Italian modernity, thereby setting the stage for Mussolini's widely-documented public persona as a model of virile masculinity. Her analysis reads Maciste's racialization in terms of emergent modes of masculinity tied to the nascent fashion industry, and thereby does not attempt to account for Maciste's racialization in terms of the film's subtext of colonial war, as this chapter does.

35 Alternatively, Greene reads the blonde-haired Cabiria (who is described by D'Annunzio in the intertitles "queen of all things white," see note 46 below) as a figure for Sicily, and her marriage to the Roman Fulvio at the film's end as the incorporation of Sicily (and by extension the entire Italian South) by the nation-state.

as a rather transparent celebration of Rome's imperial past in a way that lays the groundwork for Mussolini's cult of *romanità*. In the years leading up to Italy's Libyan campaign, proponents of Italian colonialism found in imperial Rome's dictum of "mare nostrum" a legitimating claim to the Mediterranean that would reach its apex some twenty years later during the fascist dictatorship's imperialist media blitz. As I've argued, Italy's claims to the Mediterranean were born from both an inferiority complex with regard to stronger colonial powers such as Great Britain and France and the purported fragmentation of the long-exploited and feverishly laboring Italian body politic.[36] Critically, both of these concerns found expression in racial terms: Italians risked racial degeneration or obliteration (as they were "absorbed" or "swallowed up" by other nationalities) and/or racial persecution (as their darkness—itself often a product of laboring in the fields or toiling in the coalmines—was interpreted in the Americas as an indication of Italian inferiority and a justification for lynching).[37] In such a context, the Punic Wars were a welcome font of cinematic inspiration, as they enabled contemporary Italians to "return" to a distant and secure past—when they were the victors of a race war, rather than the vanquished—in order to reclaim an otherwise profoundly unstable present.

If the film's narrative is often taken at face value as a thinly veiled apology for Italy's colonial war in Libya, likewise, the film's hero—the slave Maciste—has been described variously by *Cabiria*'s critics as "black," "African," or "Ethiopian," thereby reducing the character's artificial darkness to a transparent indicator of his fictional provenance.[38] *Cabiria* was Italy's first wide-release colonial film, and certainly one of Italy's earliest cinematic attempts to represent blackness for a mass audience. Yet the relationship between colonial regimes of representation and the racialized body of Maciste is by no means straightforward. The film's intertitles, which were embellished by D'Annunzio, refer quite plainly to the film's plot as narrating

36 As one parliamentarian put it: "[The Italo-Turkish War] was necessary for Italy's position in the Mediterranean [...] and the enterprise has reaffirmed the moral unity of Italy before the civilized world" (cited in Fuller, *Moderns Abroad* 48). We have seen how this "moral unity" also contains a racial imperative.

37 The lynching of Italians was, of course, nowhere near as widespread as the practice of lynching African Americans in the United States. I mention it here in reference to Pascoli's claim that "In America, [Italians] had become a bit like negroes; and, like negroes, every so often they were placed outside of the law and humanity, and they were lynched." Pascoli uses this as a justification for Italian colonialism in his address *La grande proletaria si è mossa* (The Great Proletarian, She Has Risen!).

38 Prolo refers to him as the "black slave" of Fulvio Axilla.

a great "race war" ("lotta delle stirpi"), or the supreme conflict of two adversarial races ("conflitto supremo di due stirpi avverse"). Yet this battle between two races does not apply to Maciste, whose darkened skin coupled with his status as hero—indeed, as the man who saves Rome, embodied by young Cabiria—operates outside the racial logic at play among the film's other characters that equates oriental- or Semitic-ness with the savage and decadent Carthaginians. One preliminary question that the film begs is: how were Italians to perceive Maciste's darkened skin in such a context? In the United States, at least one viewer, African American film reviewer J. Cogdell, was pleased by the film's positive portrayal of the "force and pride of Ethiopia" in the body of the slave-hero Maciste.[39] Italian reviewers, on the other hand, seemed to have been aware of Pagano's Genoan provenance. What were they to make, then, of his body paint and cheetah pelt? Rather than answering this question definitively through an historical reconstruction of Maciste's popular reception, my goal in the remainder of this chapter is to read Maciste's uncanny racialization in the context of the film's ideological investment in the biopolitics of colonial war.

In *Cabiria*, race operates in four distinct but interrelated ways. First, as I have mentioned, race is already inscribed—be it implicitly or explicitly—in Italy's biopolitical economy of territorial/corporeal loss/return. Next, race is staged in the film through contrasts between epidermal lightness and darkness. The third way in which race is at work in this film is in line with anti-colonial thinkers, beginning with Aimé Césaire and Frantz Fanon, who have argued that it is impossible to understand colonial relations of power without taking seriously the degree to which race served to produce, justify, and nourish the colonial imagination on both sides of the encounter. Finally, if an analysis of colonial representation requires taking stock of how race structures its myriad manifestations, following Tom Gunning and Linda Williams, I also take seriously the interrogative injunction: "how can one not write about race and the cinema?"[40] *Cabiria* thus calls out for a reading that

39 According to Lant (who raises doubts as to whether Cogdell actually saw the film), African American appraisals of early cinema in the U.S. were often concerned with the amount of time-space black characters occupied in film, and Cogdell found in Maciste an exemplary black character, one whose physical presence and psychological depth were exceptional when compared to other cinematic representations of the day (218).

40 This question is posed with regard to the U.S. in Gunning 50. Gunning ties the evolution of cinematic forms to the portrayal of race in the U.S. See also Williams, who argues that race and melodrama are two essential nodes to which the entire history of American film is bound.

is attentive to the racial and colonial dynamics at play in its narrative, visual, and ideological architecture.

To be sure, it is not only blackness that interests Pastrone—it is perhaps more apt to describe the film's overall aesthetic as clamorously Orientalist (Fig. 4.4). In addition to Maciste, several other of the film's supporting characters are emphatically racialized, from the sly Jewish innkeeper to the eroticized Carthaginian princess Sophonisba.[41] Still, to the extent that *Cabiria* may be said to participate in the larger biopolitical context that I have been discussing, the centrality of a blackened hero to this narrative of corporeal and territorial "return" requires some careful attention. If the Italian presence in Libya brought about a heightened precision with regard to Italian representations of blackness, one question that arises is whether this film itself constitutes such a rupture, or whether it can be said instead to anticipate one. In slightly different terms: does Maciste's blackness belong to a new canon of racialized representation that emerges with this unprecedented colonial conquest, or does it instead remain squarely within the biopolitical? Is the "race war" that D'Annunzio's intertitles celebrate indeed about dominating or obliterating an African other, or is it instead about improving the Italian race through territorial acquisition and corporeal restitution?

Karen Pinkus has traced Italian cultural preoccupations with blackness before fascism to F.T. Marinetti's 1909 *Mafarka le futuriste, roman africain* (translated into Italian in 1910 as *Mafarka il futurista*). For Pinkus, Mafarka, as a sexually disquieting "figure of blackness" (in spite of the fact he is coded specifically as Arab, he is distinguished from white heroes insofar as he is "colored") gave birth not only to his parthenogenetic son Gazourmah but also to a paradigm punctuated by bourgeois European concerns about licentiousness, racial mixing, and the fluidity of borders.[42] Pinkus traces the movement of what she calls the "Mafarka paradigm," a "narrative figure of the monstrous and taboo," through a representative iconography of blackness, focusing her attention on racialized bodies in advertising from the 1920s and 30s and cataloguing them under typological headings such as "The Smiling Negress," "The Silent Arab," "The *Moretto*," and so on. Though Pastrone's film follows chronologically the release of Marinetti's incendiary novel, it turns within a different representational orbit. Maciste is neither hypersexualized

41 For more on the various racial hierarchies at play in the film, see Greene.

42 On parthenogenesis in *Mafarka*, see: "Mafarka and Son: Marinetti's Homophobic Economics" in Spackman, *Fascist Virilities* 49–76.

Fig. 4.4 Sophonisba in *Cabiria*

nor subservient; he takes cover under the veil of feigned darkness (his character is covered in body paint), yet he stands in patent opposition to the blackness of blackface. In a brief but revealing scene in which Maciste and his Roman master Fulvio Axilla take respite in a Carthaginian tavern, they are served by a rotund woman who is unmistakably coded as the "mammy" figure popularized during the Civil War through Jim Crow eras in the United States (Fig. 4.5).[43]

The jubilant server smiles coquettishly as she offers Fulvio and Maciste a carafe of wine. The ivory of her eyes and teeth appear in stark contrast to her ebony-enhanced complexion, a visual and literary topos about epidermal blackness dating back to the earliest days of the European "exploration" of Africa. As she moves her shoulders to and fro in flirtatious and accommodating rapture, Maciste turns away from her, facing the film viewer as he

43 While the popularity of the minstrel show in Italy has yet to be fully explored, Jacqueline Reich speculates that its iconography was familiar to Italian filmmakers and filmgoers at the time given the widespread popularity of American films in early twentieth-century Italy. Following Gerard Butters (7), Reich notes that an overwhelming majority of American films produced in that period featured white men in blackface rather than African American actors (250). For a discussion of minstrelsy and its origins in the popular tradition of *commedia dell'arte,* see Greene 27–28.

Fig. 4.5 Fulvio Axilla and Maciste share a laugh at their adoring female server in *Cabiria*

stifles a raucous laugh by covering his mouth with his hand. Seeing him, Fulvio too erupts with laughter, pointing at the hospitable "mammy." The two share an amused drink, and the scene abruptly ends. This moment of conviviality between the patently white Fulvio and the darkened slave Maciste at the expense of the hostess seems to caution us against reading Maciste's darkening as consistent with the practice of blackface. The scene stakes out a racial boundary between Fulvio and Maciste (as laughing subjects) on one side and the mammy character (as ridiculed object) on the other. Furthermore, Maciste's and Fulvio's laughter at the hostess suggests that Maciste's comic nature is to derive from his playful strongman posturing rather than his "put on" skin color.

Maciste's racialization fails to conform to the profile of more familiar racist types—he is neither childish and meek nor licentious and aggressive; the figure would certainly not appear in Pinkus' "catalogue of abjection." Indeed, film scholar Shelleen Greene has recently added an additional interpretative layer to Maciste's racialization in the film: as a figure of the "mixed-race subject [who figures] the post-unification division of Italy into north and south [...], two racially distinct regions" (17). Greene continues, "[In] *Cabiria*, by way of the mixed-race Maciste, external colonization of Africa helped resolve [...] the racial otherness of Italy's southern populations"

(20). Another alternative to reading Maciste's darkness as a mere symptom of his other- or African-ness, would be to situate it within the film's economy of light. Fire is used both as a recurring thematic and a lighting technique, and the racialization of the film's principal characters is carried out through the visual topoi of stark contrasts between epidermal lightness and darkness. Drawing from studies of how the photographic media work through and with light to construct and reproduce whiteness, our first question is: how might *Cabiria*'s use of fire shed light on Maciste's darkness?[44]

The explosion of Mount Etna—an example of the film's use of fire as a lighting source and of its pyrotechnic monumentality—sets the story in motion, causing Cabiria's home to go up in flames and placing her in the hands of her servant Croessa. From the outset, Cabiria is associated with both fire and light—viewers learn from the typically D'Annunzian intertitle that her name "carries the genius of the industrious flame," and her flowing white gowns and the pallor of her face are foregrounded in nearly all frames in which she appears.[45] Cabiria and Croessa are captured by Phoenician pirates, who bring them to Carthage, where they are sold to the High Priest Khartalo, who selects Cabiria for sacrifice to the fire god Moloch (Fig. 4.6).

Croessa narrowly escapes imprisonment, and happens upon the Roman Fulvio Axilla, who, along with his loyal slave Maciste, is in Carthage on a

44 In his reading of the racial unconscious of American *film noir*, Eric Lott suggests that both the moral "darkness" or "shadiness" traditionally thematized within the genre and the use of chiaroscuro are structured by a racial logic ("The Whiteness of Film Noir"). Two classic studies of whiteness and the photographic media are Winston; Dyer.

45 Giovanni Pastrone allegedly paid D'Annunzio 50,000 lire for his role in *Cabiria*. Though Pastrone gave all authorial credit to D'Annunzio in advertisements and press releases for the film, D'Annunzio's actual contribution was limited to the embellishment of the intertitles, which were authored by Pastrone. In addition to adding poetic flourish to Pastrone's text, D'Annunzio is responsible for the selection of the names of (among others) the film's most celebrated protagonists, Cabiria and Maciste (Prolo 68). Critics agree that Pastrone's solicitation of D'Annunzio was an essential component of his theatrical publicity campaign for the film, which included—in addition to bombastic announcements of the film to come in major periodicals such as the *Corriere della sera*—an aerial flyover in Rome to announce the film's premier in April 1914. The plane was piloted by Giovanni Vildner, who was famous for his Trieste–Rome raid (Chimirri 95). For more on D'Annunzio's role in the making of *Cabiria*, see: Alovisio, "Il film che visse due volte"; Faccioli; Cherchi Usai, "Lettere di Giovanni Pastrone a Gabriele D'Annunzio," "Lettere di Gabriele D'Annunzio a Giovanni Pastrone," and "Lettere di Ildebrando Pizzetti a Gabriele D'Annunzio," in *Giovanni Pastrone* 71–97.

Fig. 4.6 Khartalo examines Cabiria's face before marking her forehead as "sold" in *Cabiria*

reconnaissance mission for the Roman army. Croessa enlists their help in the rescue of Cabiria from the sacrificial pyre. Maciste heroically wrests Cabiria from the flaming mouth of Moloch, and the adventures of Fulvio, Maciste, and Cabiria ensue—first comic and eventually tragic. Cabiria eventually returns into enemy hands. Viewers meet her again some ten years later as the slave Elissa, "queen of all things white and of perfect silences."[46] Maciste comes to her rescue once again, and the film's triumphant concluding scene shows Cabiria, Fulvio, and Maciste aboard a ship bound for Rome. Cabiria-the-character, insofar as she is endowed with whiteness, light, and/or fire is thus a figure for cinema.[47] This play is made evident by Pastrone's decision

46 As in the preceding chapter on D'Annunzio, my references to whiteness here are based on the language of D'Annunzio's intertitles, which refer explicitly to Elissa/Cabiria as "queen of all things white" (Chimirri 111). In Pastrone's scenario for *Cabiria* (which D'Annunzio embellished for the film's intertitles), Elissa is referred to as "the blonde slave" (Alovisio and Barbera 45–52; Alovisio).

47 A maxim of Federico Fellini's: "Films are light" (Malkiewicz 1). Cinema has long been hailed as the medium of light. In his seminal work on whiteness and the photographic media, Richard Dyer explains the racial stakes of lighting, as he argues that "a culture of light" has been inextricable from the construction of whiteness (82–144). In film studies, lighting has been studied as a technical part of filmmaking; as part of film's

(following D'Annunzio's suggestion) to change the title of his masterpiece from *Il romanzo del fuoco* to *Cabiria*—thus metonymically fusing film and protagonist (Chimirri 33; Prolo 68).

In this reductive plot summary, I have bracketed descriptions of the spectacular scenes depicting historic Roman and Carthaginian battles, though many of the film's claims to groundbreaking visual effects rest upon these scenes. The historical epic was one of the most successful genres in the "golden years" of Italian cinema, during which other significant films celebrating Roman primacy, such as Enrico Guazzoni's *Quo Vadis?* (Cines, 1913) and Giovanni Enrico Vidali's *Spartaco* (Spartacus, Pasquali & Co, 1913), also enjoyed success.[48] While the conventions of the genre and the possibilities for spectacular visual effects doubtless explain at least in part Pastrone's affection for the depiction of bellicose maneuvers in *Cabiria,* in addition, the film stages what Paul Virilio (1984) has identified as a "logistics of military perception" that emerged during World War I. Virilio situates his reading of the complicity between military modes of seeing and cinematic perception in World War I because during that war, as he puts it: "a supply of images would become the equivalent of an ammunition supply" (1).[49] Indeed, in the early years of Italian cinema, the "seventh art" was lauded by critics for its capacity to complete the task of Italian unification; the motion picture would be able to finish the work that the battles of the Risorgimento and Italy's early colonial incursions had struggled to accomplish: restoring

industrial and technological history (i.e.: the transition from gas to electric light); as the material condition for film itself (light projected onto a screen), and/or for how lighting lends depth to a film's ideological content. Classic studies of how film lighting structures or reveals ideology include: Baxter; Kracauer. For a survey of a range of studies of film and light, see the introduction to Guerin. "Light" also comes up in the philosophy of cinema, most notably within the work of Gilles Deleuze. For Deleuze, "light" encompasses the field of the visible, or the seeable (as distinct from the affirmable, or the "sayable"). On the other hand, "[For Deleuze], the cinema offers a medium in which to grasp the fluctuating relationship of the articulable and visible. [...] The cinema gradually discovers ways to proliferate disjunctions between the visible and the articulable, thereby catalyzing a kind of thought that diverges from strict determination." Cinema thus reconfigures such regimes of truth by resisting the over-determination of the image by the field of discourse (Flaxman 25–26).

48 There is some discrepancy about what constitutes Italian film's "golden years." For Prolo, they span from 1911 to 1920 (6). For Rhodes, they begin in 1908 and run through Italy's entry into World War I in 1915 (308).

49 I am grateful to Timothy Campbell for suggesting this reading and its potential implications for Pastrone's *Cabiria* to me.

Italy's status among western European powers and ensuring its territorial and biological wholeness (Rhodes).

As Virilio argues, during the years of World War I, innovations in aerial reconnaissance and surveillance photography and film were to gain ground in ways that would have lasting effects—not only on the wartime perception of the infantry and officers for whom such technologies were strategically crucial, but also on the civilian photographic media. From the darkness of the trench, the soldier's gaze illuminated his target thanks to the visual prosthetic of his rifle's telescopic sight (Bishop and Phillips). Yet, as Virilio points out, not only were soldiers "actors in a bloody conflict" that saw the replacement of hand-to-hand combat with distance fighting requiring not only optical prosthetics such as riflescopes, but also rapid-fire weapons. "[Soldiers] were also the first spectators of a pyrotechnic fairy-play whose magical, spectacular nature some of them [Ernst Junger, Apollinaire, and Marinetti] could already recognize" (87–88). In Virilio's analogy between war and cinema, the soldier who both produces and consumes the explosive spectacle of war from the cover of darkness may be considered to parallel, at varying moments, the cinematic cameraman, projectionist, and/or viewer. It is here—and not within the "transparency" of the film's imperialist aesthetics—that I would like to situate Maciste's darkness.

In fact, Pastrone's film may well be anticipating Virilio's thesis by drawing an explicit connection between wartime and cinematic technologies. In a scene depicting Rome's defeat by Carthaginian ally Syracuse, Pastrone stages Archimedes' legendary discovery of burning mirrors that are eventually able to set fire to the Roman fleet (Figs. 4.7 and 4.8). Like the projectionist, Archimedes directs his mirrors at a blank screen; the spectacle that results from his experiment in long-range assault is a flame—a flame which delights both Archimedes-as-projectionist/spectator and as pyro-strategist. Once projected onto his proto-cinematic screen, the image reveals itself as a kind of weaponry, capable of setting fire to Roman battleships from a distance. Yet, like Cabiria-the-character (a figure for cinema), Archimedes' flaming image is threatened to be consumed by the very conditions of its existence. Cabiria's name invokes her mythical origins—born from fire—and the narrative turns on her repeated capture and near sacrifice to the flaming mouth of Moloch (this scene was the one that caught the attention of the artists who designed the film's publicity posters). Fire thus both generates the cinematic spectacle and threatens to destroy it—let's not forget the missing original of *Cabiria*-the-film, rumored to have perished, like so many other reels of nitrocellulose, in a fire. Returning for a moment to Virilio, "Just as the nitrocellulose that

Figs. 4.7 and 4.8: Archimedes' burning mirrors in *Cabiria*

went into film stock was also used for the production of explosives, so the artilleryman's motto was the same as the cameraman's: lighting reveals everything" (20).

Maciste's artificial darkness therefore has something to do with his ability not only to see, but to produce, through contrast, the light that the film

Fig. 4.9 Maciste the surveying subject in *Cabiria*

thematizes through fire and the whiteness of its protagonist. It is critical to note here that Cabiria's final rescue hinges on Maciste's ability to recognize Cabiria in the slave Elissa after ten years have passed. Significantly, while Fulvio and Maciste's characters are both visibly unmarked by the passage of time, it is not Cabiria-as-Elissa who recognizes her heroic liberators from years past. It is Maciste who is given a privileged capacity for visual perception. From the first shot in which he appears, he is quite conspicuously aligned with the gaze (Fig. 4.9). While his master, the Roman Fulvio, is the official "spy," sent to Carthage on a covert reconnaissance mission, the shot in Fig. 4.9 establishes Maciste as the one who surveys and sights, while Fulvio occupies the background of the frame, fumbling somewhat impotently with his map and local guide.

Moving ahead to the final rescue scenes of the older Cabiria: Maciste and Fulvio are taken hostage, and Cabiria takes pity on the prisoners (whom, again, she does not recognize) (Fig. 4.10). Here, the composition of the shot—a blinding sunlight viewed from within the aperture in the dungeon wall—gives way to a vision of Cabiria, thus exemplifying the link between the title character and cinematic projection. In a subsequent scene, viewers again see Maciste under the cover of darkness as he spots the older Cabiria and attempts a second heroic capture (Fig. 4.11). These shots recall the architecture of cinema: Maciste occupies enclosed, quadrangular spaces,

Fig. 4.10 Elissa (Cabiria) and Maciste in *Cabiria*

Fig. 4.11 Maciste in the "projection room"

while the spectacle—Cabiria—moves past before his eyes. In both of these critical scenes that play on the recognition and rescue of the title character, Maciste is framed in spaces that might be likened to the projection room. In addition, Maciste is likened to the cinematic spectator who views the flaming filmic spectacle from a position of darkness: he is the one who spots Cabiria, and eventually enables her return to Rome. His positioning as spectator brings us back to Virilio's formulation: the soldier as both a choreographer of the spectacle of war (through his "lighting up" of the battlefield with rapid-fire guns) and as a consumer of that spectacle. Aligning Maciste-the-spectator with the soldier on the battlefield allows us to conceive of his body paint and his cheetah pelt as a kind of camouflage. Yet, like the soldier, Maciste not only consumes the scene before him: he also creates it. In both instances, he emerges from his confinement in the projection room, eventually joining Cabiria in the frame; his darkened skin thus amplifies Cabiria's spectacular whiteness. In this way, Maciste's darkness becomes part of the spectacle of the film's dazzling light play. Furthermore, Pastrone's equation of Cabiria with fire, whiteness, and light is read alongside his interest in Archimedes' burning mirrors, a further possible link between *Cabiria*-the-film, or cinema in general, and war weaponry. Critically, a black and white racial logic is already implicit in this equation, suggesting that with increasing frequency, particularly under fascism, constructing whiteness would be one of the primary cinematic battles to be waged on the Italian screen (Ben-Ghiat, "Envisioning Modernity"; *Fascist Modernities*).

How to reconcile this techno-theoretical reading of race in this film with the biopolitical context of colonial war? If, as Pastrone suggests, and as Mussolini would claim a decade later, "cinema is the [regime's] strongest weapon," and Maciste is a soldier-spectator, what kind of war are viewers watching in *Cabiria*? Unlike the films that documented the Italian invasion of Libya by Luca Comerio, the film is displaced geographically and chronologically from the event of colonial war. Furthermore, as Giorgio Bertellini has argued, the Maciste serials and other strongman films produced alongside them, include "no real violence: confrontations, duels, and battles are somewhat harmless and even cartoonish" ("Colonial Autism" 261). Bertellini's connection between these "giganti buoni" ("good giants") and the myth of "Italiani, brava gente" (Italians as good colonizers) is certainly well taken.[50] No blood is spilled in *Cabiria*; the epic battle scenes appear more to showcase the necessarily distant grandeur of the cinematic set than to recruit viewers for the bodily sacrifice of combat (as did Alberini's 1909 ode to the Garibaldian boy-soldier with which this chapter began).[51]

50 Likewise, scholars have acknowledged the "strongmen" as precursors to fascist masculinity (Renzi).

51 For Bertellini, such distant grandeur would result from the film's "antiquarian

Biopolitics reveals how the film stages war as productive rather than destructive, as life-affirming rather than life-negating.[52] It demonstrates how the film's ideological investments lie less in advocating colonial war as taking place between two adversarial races, and instead presents a version of war that is productive: lands and bodies once lost are now returned to their rightful place, and life is affirmed. Indeed, Pascoli's famed speech advocating Libyan invasion, *La grande proletaria si è mossa* (The Great Proletarian, She Has Risen!), singles out one of the heroic peasant-soldiers hailed in the speech who "is forced to bring death but wishes to bring nothing but life" (Baranello 14).[53] He continues:

> [Noi…l'Italia in guerra], combattiamo e spargiamo sangue, e in prima il nostro, non per disertare ma per coltivare, non per inselvatichire e corrompere ma per umanare e incivilire, non per asserire ma per liberare. Il fatto nostro non è quello dei Turchi. La nostra è dunque […] guerra non offensiva ma difensiva. Noi difendiamo gli uomini e il loro diritto di alimentarsi e vestirsi coi prodotti della terra da loro lavorata, contro esseri che parte della terra necessaria al genere umano tutto, sequestrano per sè […] senza coltivarla, togliendo pane, cibi, vesti, case, all'intera collettività che ne abbisogna. (18)

> [[We are an Italy at war] who fight[s] and shed[s] blood, foremost our own, not to devastate, but to cultivate; not to degenerate or to corrupt, but to humanize and civilize; not to enslave, but to liberate. Our reality is not that of the Turks. Our war is therefore a defensive act, not an offensive one, despite how our individual strategic and tactical actions may appear. We defend our men and their right to feed and clothe themselves with the products of the land worked by their own hands, against others who seize for themselves and close off, without cultivating it, land that is necessary and workable for all men, taking bread, foodstuffs, clothes and homes from the greater collective that has need of them.] (12)

solipsism," which he argues was the primary mode of Italian film from 1905 to the early 1910s ("Colonial Autism" 258–259).

52 For Hardt and Negri, in today's global biopolitical order, war is a "permanent social relation" that "brings death, but also, paradoxically, must produce life" (*Multitude* 12–13).

53 Unless otherwise noted, I have used Baranello's English translation of Pascoli's speech.

The targets of this "defensive" war of cultivation, civilization, and liberation were of course (and as we first saw in Franchetti), Italians themselves, who stood to gain the capacity to feed and clothe themselves through agricultural work. Furthermore, while the targets of defense are "uomini," or "men" (mankind), the offending group (the Ottomans) are "esseri," or "beings." This is precisely how biopolitics functions: in this case, the "men" have privileged access to rights and a political life, while the "beings" do not. This is only confirmed by the fact that Italians' status as "men" grant them entry into "the entire collectivity" (or the common), while the "beings" are presumably excluded from it and therefore subject to death.

In *Cabiria,* Pastrone picks up Pascoli's logic, which uses Roman primacy to justify Libyan conquest as a "defensive" or life-affirming war. Like Pascoli, Pastrone's narrative stages the Italian invasion of Libya as a "return" to its Roman past as an imperial Mediterranean power. This territorial return is paralleled by the corporeal return of young Cabiria, against the backdrop of what D'Annunzio's intertitles explicitly term "the supreme conflict of two adversarial races." And yet, thanks to the cartoonishness of the violence, as well as the "put on" character of Maciste's race, colonial war in this film appears to be not so much about eliminating an enemy race, but, true to biopolitical form but, instead, about improving the (Italian) race. Maciste's uncanny darkening—in addition to thematizing the capacity for the emergent medium to represent a colonial confrontation that would be played out in ever more black-and-white terms, particularly throughout the 1930s—seems also to remind us that what Italy stands to gain in Libya is ultimately self-referential. If colonial war is productive, a way of adding in territorial terms what has been "lost" in corporeal ones, Maciste's racialization serves to signal Italy's initial colonizing project not so much as a confrontation with an African "other," but instead Italy's ongoing struggle with versions of itself.[54] Bertellini and Greene have recently drawn similar conclusions, albeit through different routes, about the meaning of

54 Giorgio Bertellini's reading of Maciste's racialization as "autistic" and/or "solipsistic" arrives at a similar conclusion, though I am less interested in the necessity of the African "other" for Italian racial discourse, and have focused instead on Italian colonialism's self-referential nature within the frame of biopolitics. In addition, my reading places more emphasis upon the imagined territorial parameters of Italian racial discourse. Similarly, John David Rhodes has suggested that the monumentality of *Cabiria* stages a parallel between the conquest of cinematic space and that of colonial terrain (unpublished essay titled "Making Room for Empire: *Cabiria* and the Production of Cinematic Space" and cited in Lant 222, n. 17).

Maciste's racialized body.[55] In addition to what I described above as the technological function of Maciste's darkness (to produce, through contrast, the whiteness of Cabiria, and, following the chain of signification that I have been assembling, that of all "lost" Italians), the film's indebtedness to Pascoli's rhetoric enables us to add still more depth to our reading of Maciste's racialization. The uncanniness of Maciste's racial coding—he is (to Italian viewers, at least) at once recognizably Italian and artificially darkened—renders the racial stakes of Libyan conquest visible. It betrays a logic much like Pascoli's, according to which Italians are bound for racial persecution if they continue to emigrate toward the Americas, rather than to Libyan shores: "In America, they became a bit like the Negroes, these compatriots of the man who discovered her. And like the Negroes, they were sometimes outlawed and dehumanized; they were lynched" (8). Maciste's darkened skin makes oblique reference to these hard-laboring Italians on whom Pascoli's Libyan rhetoric depends. The working masses, figured by Pascoli with the gendered "grande proletaria," darkened by toiling in the coalmines or laboring under a harsh sun, are the rhetorical lynchpins of his famed address. With Libyan conquest, Italians thus stood only to gain, not only territorially and economically, but also biologically. Returned to their "proper" place (for Pascoli, a planned, rather than a "spontaneous," colony) and delivered from racial persecution, their blood would nourish the soil that in turn would nourish them.

Let us not forget that Bartolomeo Pagano, whose strongman posturing in *Cabiria* helped launch an international genre, was discovered by casting scouts while unloading cargo ships on the docks of Genoa.[56] Indeed, a 1914 *Corriere della sera* review of *Cabiria* declared:

> Il pubblico dei grandi e dei piccolo fraternizza per così dire con Maciste; ne ammira la superba bellezza delle forme erculee che lo fanno un raro campione della nostra razza (egli venne scoperto a Genova, dove esercitava l'umile e nobile mestiere di scaricatore nel porto), lo applaude, si addolora della sua prigionia, gode della sua liberazione, sorride con lui, distacca la

55 Shelleen Greene in her reading of Maciste as a "mixed race" Italian focuses on his figuring as a stand-in for southern Italian laborers (14–49).

56 Vittorio Martinelli constructs a narrative of physical decline around the figure of Pagano as he notes that by 1928, after an incredibly productive acting career, Pagano had retreated to his Villa Maciste in order to convalesce after bouts of paratyphoid fever and arthritis had reduced him to an incredibly thin and frail man confined to a wheelchair (10).

> sua nera figura dallo schermo e se la fa amica, portandola poi con sé nella memoria, per raccontarne le prodezze ed esaltarne la bella naturalezza dell'azione. (*Corriere della sera*, "Maciste in persona al *Lirico*")[57]
>
> [Whether young or old, the audience befriends Maciste; they admire the superb beauty of his Herculean form which makes him a rare champion of our race (he was discovered in Genoa, where he practiced the humble and noble profession of dock worker), they applaud him, feel the agony of his imprisonment, rejoice in his liberation, smile with him, detach his black figure from the screen and turn [it] into a friendly companion, always accessible in their imagination, in order to tell stories about his deeds and glorify the natural beauty of his action.] (Reich 245)

For Reich, this passage indicates that in the case of Maciste, "color is, as personified by the figure of Maciste, superceded by the national" (245). Reading the film's treatment of race in the context of Italian colonial rhetoric and biopolitics, we have uncovered an even more nuanced understanding of this critic's notion of "our race." Reading symptomatically this reviewer's conflation of the character Maciste and the actor Pagano, one finds an example of how Maciste is to be consumed visually as belonging not to some other, "African race," but to "[the Italian] race" on the basis of his once laboring (and now comically mighty) body. Indeed, in describing the widespread appeal Pagano's Maciste had to bourgeois and working-class audiences alike, Luca Cottini writes that one of the reasons for his popularity was precisely his comic "tendency to solve problems by using the strength of his arms." Furthermore, for Cottini, what granted Pagano's Maciste such lasting popularity in the wake of *Cabiria* was precisely his recognizability as the embodiment of the ideal Italian man.

Fusing the Symbolic Landscape

Biopolitics enables us to conceive of Italian racial discourse before fascism in "productive" terms, or at the very least, in terms of an economy of loss and restoration and/or return, rather than simply an economy of racial superiority and inferiority. Since well before the culminating battles of

57 I am grateful to Jacqueline Reich for providing me the original reference and source.

the Risorgimento, Italy experienced its national "*re*birth" as the end of an enduring occupation by foreign (primarily French and Austrian) oppressors; as we have seen, this sense of inferiority with regard to Europe, as well as the particularly Italian experience of emigration, became critical components of early colonial rhetoric and the rhetorics of loss that subtended it, and later, of D'Annunzio's imperialist poetics at Fiume. By way of conclusion, I'd like to turn for a moment to two films made in the wake of *Cabiria* that illustrate the persistence and evolution of some of the themes I have been discussing, particularly: Maciste's racialization in light of the film's conflation of territorial and corporeal restitution, and the black, white, and green of Libyan conquest, or, the inextricability of agricultural rhetoric from Italian racial discourse. In addition, both films continue where *Cabiria* left off, the first, Pastrone's *Maciste alpino* (Maciste the Alpine Soldier, Itala Film, 1916), by following Maciste to contested northeastern lands, and the second, Carmine Gallone's *Scipione l'Africano* (Scipio the African, Ente Nazionale Industrie Cinematografiche, 1937), by narrating the Punic Wars from the perspective of Scipio Africanus (who plays only a minor role in *Cabiria*). *Maciste alpino* was filmed during World War I and marked lift-off in Maciste's (Pagano's) ascent to serial celebrity as a "strong (white) man." *Scipione l'Africano*, produced by the regime at the height of its racist mania (sandwiched chronologically between the 1935 invasion of Ethiopia and the drafting of the racial laws in 1938) takes another cinematic stab at the Punic Wars with quite a different approach to racialization than *Cabiria*.

In my discussion of *Cabiria* above, I focused upon the film's rhetorical investment in Rome's territorial "loss" of Mediterranean Africa by following Maciste, Cabiria, and Fulvio through their adventures in Carthage. One of the scenes I did not discuss, shot on location in the Graian Alps of Piedmont, depicts Hannibal's historic crossing of the Alps. Back in Carthage, Fulvio receives word of the "impending danger for his far-away fatherland [*patria*]," and quickly retreats for Rome. (This is where *Scipione l'Africano* will pick up the story in 1937, as it begins with Hannibal's oppressive Roman reign). In addition to staging a territorial loss, *Cabiria* thus also stages a territorial breach whose contemporary correlative is to be found in the "unredeemed" lands of Trieste, Trent, and the South Tyrol held by Austria-Hungary (the basis for these claims lay, of course, in the peoples who inhabited these lands, whom *irredentisti* such as D'Annunzio claimed were not only linguistically but racially Italian). This is precisely where *Maciste alpino*, the far lesser known follow-up to the hugely successful *Cabiria*, takes place. The film stages yet another gendered corporeal loss—the daughter of a Count whom Maciste has befriended—and the plot

Fig. 4.12 *Maciste alpino* (Itala Film, 1916)

is driven by Maciste's capricious yet lighthearted feats of strength against Austrian soldiers in his attempts to rescue the maiden. Critically, by *Maciste alpino,* Maciste has shed his blackened skin and instead dons the unmistakable feathered cap of the Alpine soldier (Fig. 4.12).

If we were to map the geographies of these first two Maciste films alongside how they stage the racialization of their hero, we have one celebrating the return of Italy's Mediterranean "fourth shore" and featuring a blackened hero (whose uncanny racialization serves more to underscore Italian "whiteness" than to represent any consistent figure of "blackness"), followed directly by a film whose subtext addresses Italy's struggle with the Austro-Hungarian Empire over the "unredeemed" Italian territories. Maciste's transition to unequivocal whiteness (recall the "rosy-cheeked and solemn Alpine soldier" in contrast to the "the dark and slender Sicilian" of Pascolian colonial poetics) thus occurs alongside a therapeutic circumscription of "missing" Italian territories to the Northeast and the South—remember Pascoli's claim in *La grande proletaria,* "Earth, sea and sky, alps and plains, peninsula and islands, North and South are *perfectly fused*" (Baranello 11, emphasis mine), as well as D'Annunzio's aeronautical vision that seamlessly linked not only Libya to Eritrea and Somalia, but also to the battlefields of World War I and the unredeemed territories of Istria (including Fiume). That cinema has played a pivotal role in evoking a symbolic landscape that is either national or colonial (though, as I have been arguing, one is never entirely immune from the other) is clear, then, from the earliest examples of Italian film, which interpellated

Fig. 4.13 Scipio (white horse) vs. Hannibal (black horse). Gallone, *Scipione l'Africano* (1937)

their viewers through visual (and invisible) logics of territorial and corporeal wounds and healing.[58]

Shifting our focus two decades forward to the propaganda film *Scipione l'Africano*, we have a significant refashioning of the racial logics at play in *Cabiria*. Mussolini's obsession with *romanità* finds full expression in this expensive regime-financed cinematic colossus that narrates Scipio's triumphant departure for Carthage and his battle with Hannibal in the North African desert. Produced during the regime's Ethiopian campaign, *Scipione*—like *Cabiria*—stages an analogy between past and present through an even more dubious geographic displacement: ancient Carthage stands in for modern Ethiopia, as the clash is referred to as occurring between "Italy" and "Africa." The film thus depicts the battle between Rome and Carthage in brashly black and white terms: during the epic Battle of Zama scenes, Hannibal (who is portrayed in the film following standard iconography: with dark skin and hair, a curly beard and an eye patch), flanked by black African extras dressed as Carthaginian soldiers, rides a black horse, while Scipione rides a starkly white one (Fig. 4.13).

The drama of these climactic battle scenes is heightened through repeated cuts between reciprocally charging black and white horses. The film opens

58 On how symbolic and material landscapes inform cinematic practice, and vice-versa, see Harper and Rayner. See also Denis Cosgrove's seminal study.

with a shot of a battlefield full of Roman war dead and an exhortation to "Avenge [*vendicare*] the dead of Canne," which would have resonated at the time in Italian ears with pro-colonial cries to "Avenge the dead of Adwa." Its conclusion closes the chain of signification with which it began: the final scenes begin with a battlefield full of Carthaginian war dead, as a Roman soldier rises and proclaims, "The dead of Canne are avenged! Hannibal is defeated!" Voices ring out over victorious legions boarding ships returning to Rome. "Carthage is destroyed! Rome is saved!" The final scene depicts Scipio at home once again, surrounded by his servants, wife, and children. Scipio appears in the foreground (his wife and infant are visible in the background), and dips his hand into a bushel of grain, solemnly proclaiming, "The grain is good, and tomorrow, with the help of the gods, we will plant it" ("Buon grano. E domani, con l'aiuto degli dei, comincerà la semina.") He turns to join his family in the shot, and his young son enters the frame, approaching his father for an affectionate pat on the head. Reclamation, in at least two senses—in terms of avenging bodily and territorial loss, as well as in terms of the regime's plans for agricultural self-sufficiency as exemplified by their well-funded projects of *bonifica*—thus becomes the node around which the film's ideo-logic turns.[59] The negation of life, or avenging the dead of Canne (Adwa) by sacrificing Romans, killing Carthaginians, and wresting their land from Hannibal's grip is thus a necessary counterpart to the productive, or life-affirming, agricultural project of *bonifica*. From the play between whiteness and darkness in *Cabiria* over a backdrop of ancient loss and gain, the regime's film picks up where *Cabiria* left off not only thematically (by following another narrative thread through the Punic Wars), but also ideologically. In this later period, the inextricability of life from politics would give way to an ever more explicit subjection of the former to the latter, thereby increasing mechanisms of colonial drive and activating immunitary apparatuses in ways that are still emerging from the shadows and coming to light.

59 The film was shot on location in the "reclaimed" lands of the Pontine Marshes, illustrating the convergence between internal and external colonization during the regime (Caprotti, "Scipio Africanus"; *Mussolini's Cities*). On the fascist new towns, see: Fuller, "Tradition as a Means to the End of Tradition" 178; "Wherever You Go, There You Are." David Horn ties *bonifica* to both human bodies (i.e. the regime's demographic policies) and to agricultural new towns, which the regime cast as a fertile alternative to the sterile city (*Social Bodies* 95–122).

Epilogue

This book began with a discussion of how Italian cultural production during the liberal period and scholarly interpretations of it have either attempted or neglected to confront racialization as a critical part of the discursive formulation of Italians as modern political subjects. Subsequent chapters illustrated how biopolitics opens up the interpretative field, allowing readers to "see" race at the intersection of a variety of problem areas that preoccupied post-Unification thinkers. In calling for a biopolitical reading of Italian racial discourse, I have been taking implicit aim at two commonplaces in studies of modern Italy: the first concerns the origins of Italian state racism and the second concerns the ideological and rhetorical splitting of projects of nation- and empire-building. First, many genealogies of Italian state racism either explicitly or implicitly figure its spontaneous inception within the fascist "parenthesis."[1] This positioning is often accompanied by a(nother) narrative of Italian belatedness, in which Nazi Germany is figured as the sinister inventor of state racism and fascist Italy is depicted as merely having jumped on board an ideological train that was already in motion.[2] Such perspectives grew out

1 In 1994, as a response to this ubiquitous narrative, scholars from the Departments of History and Philosophy at the University of Bologna began a series of genealogical studies of Italian racism, organizing their pursuits under the heading: "Seminario permanente per la storia del razzismo italiano" (Burgio and Casali).

2 By 1945, as she was drafting what would become *The Origins of Totalitarianism*, Hannah Arendt had already dismissed this conventional appraisal, though few seemed to have heeded her precocious insight. "If race-thinking were a German invention, as it has sometimes been asserted, then 'German thinking' (whatever that may be) was victorious in many parts of the spiritual world long before the Nazis started their ill-fated attempt at world conquest. Hitlerism exercised its strong international and inter-European appeal during the thirties because racism, although a state doctrine only in Germany, had been a powerful trend in public opinion everywhere. [...] The historical truth of the matter is that race-thinking, with its roots deep in the eighteenth century, emerged

of political necessity. At the close of World War II, anti-fascist intellectuals who had been silenced during the fascist *ventennio* (those, that is, who did not perish in prison or exile) began to emerge from the figural and literal wreckage in order to ask how such a political tragedy could have found in Italians such an accommodating cast of characters. Historical, political, and even literary studies produced in this climate seemed to have a choice among a finite number of explanations for the rise of fascism: their answers ranged from trivialization to condemnation, from casting blame on others to parodic self-loathing. And one thing was sure: fascism, and all of its familiar and bulky apparatuses, had to be purged not only from public offices, but from public consciousness.[3] The racist persecution of Italian Jews (not to mention the colonized) was for many one of the most horrifying expressions of fascist violence, and as such, some intellectuals were eager to salvage the remnants of the Italian liberal democracy, distancing it from its nefarious successor. It was in this context that Italy's preeminent historical materialist philosopher Benedetto Croce famously proclaimed fascist Italy as a "parenthesis" in Italian history. Critical to this project of distancing was the scapegoating of Nazi Germany, particularly when it came to state racism, and thus a depiction of Mussolini as a reluctant racist, a second-rate copycat (e.g. Spinosa and Perfetti). Whether there may be a grain of truth in such accounts is less of interest than identifying what sorts of new silences or blurred vision such interpretations provoked. One explanation that gained significant ground and that continues to shadow how scholars have approached Italy's relationship to race thinking, was that Italian nationalism was primarily and inherently voluntaristic, and thus founded upon ideals of patriotic choice, of a decidedly social rather than biological contract. In contrast, German nationalism was from its inception organicistic, rooted in ideals of blood belonging and natural territory, and thus

simultaneously in all Western countries during the nineteenth century." Arendt's rich discussion proceeds, like those of Morrison and Gates mentioned below, by addressing the emergence of race thinking and racism in France, Germany, and England, though it regrettably makes no mention of these processes in Italy or Spain. See "Race-Thinking Before Racism," in Arendt's classic study *The Origins of Totalitarianism* (158).

3 The scholarship on Italian memories of fascism and the World Wars is immense. One recent study in English by John Foot examines the polarization of memory in Italy from World War I to today, in particular around traumatic events such as war and terrorism, but also extending to other areas of cultural life. Giovanni Contini, Luisa Passerini (*Fascism in Popular Memory*; "Memories of Resistance"), and Alessandro Portelli (*L'ordine è già stato eseguito*; *The Order Has Been Carried Out*) are three of the most active and prolific archeologists of memory working in Italy.

by definition more amenable to a genocidal politics.[4] This is not to say that there are not significant differences between the two nationalisms, nor that the Italian state has always been covertly racist, but instead, that the historical and political exigency that gave rise to such interpretations should be taken into account, and that the implications of such interpretations should be critically reexamined for the way in which they construct what Barbara Spackman has called the "black box" of fascism. Spackman writes: "[F]ascism is represented as a black box whose contents are unspecified but whose moral significance is given in advance" (*Fascist Virilities* 116). According to the logic that structures the binding of (immoral) racism to the (immoral) fascist state, if Italian state racism did not precede the fascist state and if, even better, fascist racism can be chalked up to junior Nazism, all can then be uncritically dumped into the trash bin of fascist aberration. Rather than a repositioning of an explicitly racist state discourse at an earlier point on a chronological plane (although this may prove to be a side-effect), this book's chapters have attempted instead a sustained analysis of what Foucault has called the "polyvalent mobility" of racial discourse across a variety of fields.[5] For Foucault, identifying a point of origin in racial discourse is impossible and futile; instead, by insisting on its adaptability and the diversity of its articulations, Foucault emphasizes the processes by which it has gained authority, and continues to do so, in diverse moments and contexts.

Another tendency in how scholars have tended to approach modern Italy has been to concentrate on processes of nation building, while holding Italian colonialism in reserve as an epiphenomenon, or an afterthought. The risk of such an approach is that in order to do this, the Italian nation-state is presented as a consolidated juridical and social body that, having achieved a degree of

4 See Federico Chabod, *L'idea di nazione* (Rome; Bari: Laterza, 1974) 68 (discussed and cited in Banti, *La nazione del Risorgimento* 56). Banti challenges this reading of the Italian nation as essentially voluntaristic by pointing out that in order to willingly submit itself to this social contract, the (organic) community must already have been formed. In his survey of the Risorgimento poetic and political canon, he argues that an Italian community was already formed on the "natural" bases of divine ordination—such is the case in the writings of Giuseppe Mazzini and Vincenzo Gioberti, for example—and/or blood belonging (63).

5 Stoler writes: "We need to understand that racial discourses, like those of the nation, have derived force from a 'polyvalent mobility,' from the density of the discourses they harness, from the multiple economic interests they serve, from the subjugated knowledges they contain, from the sedimented forms of knowledge they bring into play" (*Race and the Education of Desire* 204).

territorial and/or administrative unity, then turns its gaze outward in order to absorb additional territory. Periodization insulates the humble, liberal-democratic ideals of the Risorgimento from the raucous and less palatable "imperial age," even in the absence of a fascist alibi. The nation-state forged by the Risorgimento is thus distanced temporally and ideologically from the project of empire. This narrative recurs throughout modern Italian historiography, and it erroneously depicts colonial enterprise as a sort of prosthesis, rather than as integral to the ideological fantasy of the unified national body.[6] This disposition is typified in Jared Becker's article on D'Annunzio's orientalism, as Becker charges that legible within *Maia* is D'Annunzio's "[return] to the model of Carduccian civic poetry" wherein "he expands its scope from modest nation-building to a much more grandiose dream of empire" ("D'Annunzio, Orientalism and Imperialism" 1–2). Here, the liberal nation-state is figured in opposition (inasmuch as it is staged as a precursor) to the imperial nation-state. Putting pressure on the notion of a phantasmagorical shift from the "modesty" of liberal nationalism to the "grandiosity" of empire has been the task of this book by keeping in mind, as Miguel Mellino puts it, the "underlying coloniality" of various modes of Italian nation formation (87).

For over a quarter century, an increasingly vast field of postcolonial scholarship has begun to explore more fully the extent to which the great European nation-states and their colonies were shaped by their imperial encounters.[7] Though the Italian case is an admittedly "minor" example (although the Introduction explored the risks of such a designation),

6 The minimization of the role that Italian colonialism (as a set of practices) played in Italy's consolidation as a nation-state is characteristic of many histories of modern Italy. For a few examples of the splitting of liberal nationalism from projects of empire, see: Croce, *Storia d'Italia*; Banti, *Il Risorgimento italiano*; Davis, *Italy in the Nineteenth Century*; Hearder, *Italy in the Age of the Risorgimento*. An exception to this historiographical trend can be found in De Bernardi and Ganapini. This omission may be due at least in part to the lack of access (until the 1960s and 1970s) to archival materials, inhibiting scholars in their research of Italy's colonial past. For an account of this phenomenon, see the introduction to Ben-Ghiat and Fuller. For examples of contemporary scholarship that take the reciprocal enunciation of nationalist and imperialist projects as a point of departure, see: Ben-Ghiat, *Fascist Modernities*; Fuller, *Moderns Abroad*.

7 For a general introduction to the field of postcolonial studies, see: Ashcroft, Griffiths, and Tiffin; and Young, *Postcolonialism*. Foundational texts include Edward Said, *Orientalism*; and Ranajit Guha and Gayatri Spivak, *Selected Subaltern Studies*. On the pitfalls of the *Subaltern Studies*' critique of Eurocentric universalism, see Chibber. The formal introduction of the field of "postcolonial Italian studies" to the English reading

or perhaps precisely because of this reason, it stands to offer something important about comparative European modernities and colonialisms. While perhaps ever less frequently today (as Turkey and Greece take its place), Italy has long been hailed as Europe's "internal other" (Van den Abbeele). From its depiction by northern Europeans during the Grand Tour as a land of romantically decaying excess to its enduring representation as either a locus of spiritual and sensual reawakening (of the *Under the Tuscan Sun* or *Eat, Pray, Love* variety) and/or a beautiful landscape teeming with corrupt (and/or inept) politicians and *mafiosi,* the Italian nation-state has always been perched somewhat awkwardly at the geographic and symbolic threshold of Europe and Africa. Its liminal status stands to help us to nuance understandings of racial representation as merely the ideological tools of the dominant, as Italy has long suffered an inferiority complex, and never more vociferously than when it came to what it figured as territorial dispossession (the "unredeemed" and/or "lost" lands of the Roman Empire) in the years between its coming into being as a modern nation-state and World War I.[8]

In post-Unification Italian racial discourse, rhetorics of territorial and corporeal loss are used fetishistically to discursively "mend" a fundamental absence (Stewart-Steinberg) or constitutive fracture (Esposito) in the modern Italian subject. These textual mechanisms of disavowal—the texts under consideration 'know' very well that there is no unitary, modern, racialized Italian (or "vital") subject, but all the same they 'behave' as if there is—bring up larger questions about collective memory that Dominick LaCapra identifies as a conflation between loss and absence in the context of historical trauma ("Trauma, Absence, Loss"). Whereas losses are the result of traumatic historical events, and are therefore amenable to resolution, or working through, absence is transhistorical and constitutive, and is therefore difficult or impossible to resolve. A conflation between loss and absence lies at the heart of much of the so-called liberal Italian racial discourse we have been analyzing, and arguably has implications for its ongoing effects today. The rhetorical deployment of loss as a means of disavowing a constitutive absence causes something nefarious to emerge: "Paradise lost could be regained, at least at the end of time. One might ask," speculates LaCapra, "whether the conversion of absence into loss is essential to all fundamentalisms or

public occurred in 2012, with the publication of Cristina Lombardi-Diop and Caterina Romeo's volume *Postcolonial Italy.*

8 For a recent analysis in English of the derogatoriness of Italian national character, see Patriarca, *Italian Vices.*

foundational philosophies" (702). One may therefore detect in these lines, as well as in the pages we have been reading, the specter of fascist racial discourse.

As Stewart-Steinberg claims in *The Pinocchio Effect*, and as the texts under consideration in this book have further demonstrated, anxiety characterized the post-Unification moment and was, far from an impediment, constitutive of Italian modernity. "T]he formulation of an Italian national self was predicated on a language that posited marginalization and powerlessness as fundamental aspects of what it meant to be modern Italians," writes Stewart-Steinberg (2). For LaCapra, anxiety, "the elusive experience or affect related to absence," often leads to the identification of a specific thing or object to be feared, enabling the potential for a mastery of that fear. As LaCapra suggests:

> The conversion of absence into loss gives anxiety an identifiable object—the lost object—and generates the hope that anxiety may be eliminated or overcome. By contrast, the anxiety attendant upon absence may never be entirely eliminated or overcome but must be lived in various ways. [...] Avoidance of this anxiety is one basis for the typical projection of blame for a putative loss onto identifiable others, thereby inviting the generation of scapegoating or sacrificial scenarios. In converting absence into loss, one assumes that there was (or at least could be) some original unity, wholeness, security, or identity which others have ruined, polluted, or contaminated and thus made 'us' lose. (707)

And yet, as Esposito cautions, the very grounds upon which the human community is posited are tenuous, as community has long been both threatened and subtended by its inverse: immunity. The fetishistic mechanisms that conflate loss and absence, along with the immunological logics that I have argued shaped Italian racial discourse in post-Unification Italy may thus be read as providing fertile rhetorical terrain for the emergence of fascist racial discourse. As I will argue by way of conclusion, this conflation of loss and absence also helps to explain why scholarly and public debate about Italy's imbricated histories, race thinking, and colonialism remained for many years and until quite recently in relative obscurity.

Race Critical Italy

> The great force of feeling and imagination needs much nourishment, living aid, the sustenance of real things.
>
> Giacomo Leopardi, "Discorso sopra lo stato presente dei costumi degl'italiani" (672)

> The specificity of modern racism, or what gives it its specificity, is not bound up with mentalities, ideologies, or the lies of power. It is bound up with the technique of power, with the technology of power.
>
> Michel Foucault, *Society Must Be Defended* (258)

Notwithstanding the proliferation of racial representation throughout turn-of-the-century Italian literature, anthropology, political discourse, and visual culture (to say nothing of its preponderance during the fascist era), two somewhat recent theoretical reflections on race and literature in the United States include the same conspicuous and provocative omission: in passing references to scholarship on race within the national literatures of Europe, Toni Morrison (7) and Henry Louis Gates, Jr. (3) name nearly all western European national literatures but that of Italy. This exclusion of Italy from the pantheon of racializing European literatures presents a variety of interpretive possibilities. Is Italy's literary history not sufficiently "national," a question that has occupied authors and critics alike since Dante? Does Italy's paltry position in the colonial contest, or its relatively late experience of immigration, erase the relevance of race to its literary history? What, then, do we make of the fact that the process of national canonization in Italy has enshrined the works of not only Dante, Petrarch, and Boccaccio, but also Tasso, Ariosto, and perhaps less markedly Basile, none of whose masterpieces would be possible without the presence—shadowy or thunderous—of physiognomic, chromatic, and/or physiological, in short racial, difference? The refusal of Italy's pertinence to Europe's history of racial representation enacted by these two preeminent thinkers paradoxically and, we might safely imagine, quite inadvertently, reinforces the very rhetoric of absence that has long diverted the gaze of critics from the politics of race in Italy.[9]

9 It is of note that in the Italian context, the very expression "politics of race" has

My goal here is not to bemoan the marginalization of poor little *Italietta* vis-à-vis her more powerful European neighbors (a move which would risk recalling many justifications for Italian colonialism) as much as it is to call attention to rhetorics of absence irrelevance, and/or minor status that threaten to obscure and interrupt the way scholarship on modern Italy approaches the structures of racialization that shape Italy's literary canon, as well as the vast political and cultural landscapes on which Italian modernity has successfully or otherwise attempted to ground itself.[10]

I was motivated to write this book in part to respond to the fact that the textual production of race in Italy had, it seemed to me when I began writing, too often and quite perilously been either underemphasized for its relatively negligible impact when compared with American and other western European traditions, or dismissed as the clumsily racist stuff of an emphatically bygone (fascist) era. This relative lack of emphasis on racial thinking in Italy, resulting from the appraisals mentioned above (not unequivocally "national" enough? Not big enough of a player at the colonial conference table? Not racist enough?), is paradoxically inscribed within a colonial logic: race is more or less relevant to a given national context based upon the size or greatness of its (ostensibly former) empire. In pursuing this line of inquiry, it became clear that, far from a marginal comment in the annals of Italian history, race has been an enduring and powerful idea for generations of Italians, and it is deeply enmeshed with its history as what historian Mark Choate has quite effectively dubbed an "emigrant nation," a nation constituted at least as much by its emigrants as by those residing within the borders of the nation-state. What's more, precisely *because* of its perceived irrelevance or absence from public space and discourse (again, until the arrival of increasing numbers of racially marked immigrants beginning in the 1980s), race thinking seems to have enjoyed a relatively extended post-colonial afterlife in Italy. For evidence of this, one need not look much further than the feverishly publicized and suspiciously well-wrought "gaffes" of former Prime Minister Silvio Berlusconi (about U.S. President Barack Obama's "suntan" and the "superiority of Western civilization"). Indeed, Berlusconi's "jokes" may well express a collective amnesia or a repression

fascist connotations, as *la politica della razza* was a euphemism for the violent and persecutory racial laws of 1938–1939.

10 Here, I use "minor" in its common sense usage, rather than in the revolutionary sense theorized by Deleuze and Guattari. For a discussion of their minor literature in relation to Italy, see Parati 54–103.

of Italy's colonial past (Freud, *Jokes and their Relation to the Unconscious*).[11] Cristina Lombardi-Diop has argued that the implications of this absence or amnesia have shaped contemporary Italy as a "postracial" society, "where widespread racism permeates the political discourse, the societal behavior, and popular culture, yet race is often unnamed and ultimately silenced" ("Postracial/Postcolonial Italy" 175). Far from marking the eclipse or "overcoming" of racial discourse, "postracial Italy" refers instead to a subtle yet omnipresent racial discourse that underpins contemporary social relations in Italy. Similarly, Caterina Romeo conceives of the exclusion of race from contemporary Italian cultural debates not in terms of repression or amnesia, but in terms of an "evaporation," a discourse that is temporarily invisible, though nevertheless pervasive, and always bound to reappear.

The terms of contemporary public discourse on race in Italy suggested by Lombardi-Diop and Romeo, which resulted for many decades in a dearth of scholarly attention to race thinking in Italy, illustrate another important effect of the conflation of loss and absence that LaCapra argues produces "dubious results." In cases in which historical losses (of colonies in Libya and Eritrea, for instance) are conflated with absence (as a result of the collective amnesia about or repression of Italy's racialized colonial encounters), there emerges, "a tendency to avoid addressing historical problems, including losses, in sufficiently specific terms or to enshroud, perhaps even etherealize, them in a generalized discourse of absence" ("Trauma, Absence, Loss" 700). Still, LaCapra continues, "something of the past always remains, if only as a haunting presence or revenant" (700). While historical losses and/or traumas may be redressed by acting out or working through, when loss is generalized as (transhistorical) absence, "one remains possessed or haunted by the past, whose ghosts and shrouds resist distinctions (such as that between absence and loss)" (699). Such is precisely the condition of the "postracial" Italian "evaporation" of racial discourse theorized by Lombardi-Diop and Romeo. The conflation of loss and absence that I argue enables the ideological fantasy of the unified, racialized Italian national body thus to some degree accounts for the unresolved, spectral nature of contemporary Italy's racial politics.

Since the 1990s, scholarship on modern Italy has dealt increasingly with race, some of it in ways that are indebted to the theoretical foundations

11 For a similar reading of Berlusconi's race jokes, see Lombardi-Diop, "Postracial/Postcolonial Italy" 175. Several preeminent scholars of memory in modern Italy have used psychoanalytical models of either amnesia or repression to describe Italy's relationship to colonialism and World War II. See: Fuller, *Moderns Abroad*; Marcus; Triulzi.

laid by, among others, Morrison and Gates (their oversight of Italy's relevance notwithstanding). In both the Anglo-American and Italian academies, a critical interest in genealogies of race thinking emerged in response to the complicity of the fascist dictatorship in explicitly racialized violence (exemplified in fascist demographic policy, inaugurated as early as Mussolini's "Ascension Day Speech" in 1927; the invasion of Ethiopia from 1935 to 1936; and the persecutory racial laws of 1938–1939), leading up to what has been figured as its apex in World War II.[12] Conventional scholarly approaches to race in Italy have thereby been primarily anti-fascist, and by extension anti-racist. Viewing race from the postwar perspective of anti-fascism, several scholars widened the terms of the discussion by aiming their inquiries not only at the explicitly racist texts of the fascist era, but at a larger constellation of problems that have preoccupied Italy since well before its albeit tentative entrance into modern nationhood in 1861. A number of works published in English beginning in the mid-1990s reoriented scholarly approaches to Italian fascism by complicating the ideological rigidity that characterized conventional studies of the period by taking into consideration a range of representational practices and theoretical approaches.[13] Other important work that engages race in modern Italy has converged around three main areas: the so-called southern question, an ongoing debate circulating in a body of texts ranging from the racial scientific to the poetic aimed at formulating a resolution to the historic dissymmetry between Italy's northern and southern regions in patterns of liberal-democratic/capitalist development; migration to and from Italy; and Italian colonialism. The methodological approaches in each of these three broad fields vary, though generally speaking they all analyze race in terms of objectification, difference, and/or hierarchy. For instance, influential studies of Italy's southern question by Jane Schneider, Nelson Moe, and John Dickie address race vis-à-vis Said's Orientalism and the stereotype.[14] Recent work on

12 Barbara Spackman offers a reading of Mussolini's reproductive politics and suggests that it is already formulated in the "Discorso dell'ascensione" of 1927. See "Fascism as Discursive Regime," in Spackman, *Fascist Virilities*. Several recent studies in Italian target fascist racism (Speciale; Riccardo et al.; Cuomo; Pisanty and Bonafé; Germinario; Collotti; Israel and Nastasi).

13 For studies of fascism that address the relevance of race via fascist approaches to the body, advertising, and spectacle, see: Ben-Ghiat, *Fascist Modernities*; Falasca-Zamponi; Spackman, *Fascist Virilities*; Pinkus.

14 See Moe, *The View from Vesuvius*; Dickie, *Darkest Italy*; Schneider; Verdicchio, "Introduction"; Teti.

migration both to and from Italy by Mark Choate, Donna Gabaccia, Graziella Parati, and Pasquale Verdicchio, among others, also deals either obliquely or explicitly with race, often through the analytic of diaspora studies.[15] Finally, an ever-growing number of scholars of Italian colonialism—from pioneering historians Angelo Del Boca, Giorgio Rochat, and Nicola Labanca to cultural theorists and historians informed by postcolonial studies such as Cristina Lombardi-Diop, Derek Duncan, Jacqueline Andall, Mia Fuller, Giulia Barrera, Patrizia Palumbo, and Ruth Ben-Ghiat—have explored how the construction of racial and gender hierarchies was necessary to the subjection and rule of the colonized, with a particular emphasis on these processes under fascism.[16] One recent history has addressed the convergence of the southern question, emigration, and colonialism in an analysis of racial thought in Italy before fascism: Aliza Wong's *Race and the Nation in Liberal Italy, 1861–1911: Meridionalism, Empire, and Diaspora* (2006). Wong's book traces the origins of Italian racial discourse, which she understands as an ethnocentric mode of producing difference or "othering," to the language of liberal Italy's southern question. She argues that the metaphors and topoi of southern question discourse shaped fields such as racial science (including physiognomy and criminology), colonialism, and Italian emigration after Unification. The result, Wong suggests, is that "the lexicon of the southern question becomes the most familiar, most accessible idiom with which to discuss [these] other discourses of difference" (5).

Within these influential studies of how the southern question, emigration, and colonialism shaped nation building in modern Italy, one of the underlying premises is that nationalist discourse relies upon various models of ethnocentrism or difference. By employing a biopolitical frame, I have been posing a somewhat different theoretical question: how do rhetorics of loss (both territorial and corporeal) function fetishistically to heal or resolve a constitutive absence in the modern Italian racial subject? Similarly, in Foucauldian terms, how does discourse that purports to "make live" concomitantly "let die"? Finally, rephrased in Esposito's immunological language, how do these rhetorics that appear on the surface to affirm and safeguard the Italian national community wind up negating it from within? In order to

15 See Passerini, *Women Migrants from East to West*; Giordano; Parati; Guglielmo and Salerno; Gnisci; Clò and Fiore; Gabaccia; Dal Lago; Verdicchio, *Bound by Distance*.
16 For a representative sample in English, see: Duncan and Andall, *National Belongings*; *Italian Colonialism*; Ben-Ghiat and Fuller; Palumbo; Barrera, "Colonial Affairs"; Matteo.

attempt an answer, I have taken a microscopic approach to racial discourse not in order to rehearse or deconstruct its (erroneous) social-scientific bases, nor to chart the breadth of its articulations across turn-of-the-twentieth-century Italian culture, but instead to explore how racialized Italian subjects are produced in the languages of post-Unification nationalism through biopolitical rhetorics of (re)productivity.

How does this intersection between race and (re)productivity that I have been discussing distinguish Italian racial discourse from other European racializing traditions? In Hannah Arendt's seminal analysis of the origins of European racism, what set French race thinking apart from German or English versions was that in France, racial discourse grew from a struggle between the aristocracy and the bourgeoisie and nurtured a civil war (which meant, for Arendt, that in France, racism was not coterminous with nationalism, but was instead "antinational"). In the case of Germany, she writes, race thinking served to fuse a fragmented national population against foreign oppression:[17] "In contrast to the French brand of race-thinking as a weapon for civil war and for splitting the nation, German race-thinking was invented in an effort to unite all people against foreign domination" (*The Origins of Totalitarianism* 166). Race was also linked closely to nationalism in England, yet for somewhat opposite reasons. Rather than being used in a rhetoric of struggle against a tyrannous outside force, Arendt argues that English race thinking was tied to the overtly hierarchical structure of English nationalism: "[I]nequality belonged to the English national character" (175). As such, one of the building blocks of English society was, for Arendt, the inheritance of land and, with it, rights. It was within this discourse that race thinking found fertile ground: "The concept of inheritance, applied to the very nature of liberty, has been the ideological basis from which English nationalism received its curious touch of race-feeling ever since the French Revolution" (176). In this schema, the English people constituted "the nobility among nations": "[T]he concept of inheritance was accepted [from feudalism] almost unchanged and applied to the entire British 'stock.' The consequence of this assimilation of noble standards was that the English brand of race-thinking was almost obsessed with inheritance theories and their modern equivalent, eugenics" (176).

Though, like Morrison and Gates, Arendt does not mention Italy in her analysis of European race thinking, her parameters with regard to the French,

17 Both Arendt and Foucault trace the origins of, for the former antinational racism, and for the latter, sovereignty's appropriation of a war between the races, to the early eighteenth-century writings of French nobleman Comte de Boulainvilliers.

English, and German contexts provide us with some important points of comparison. Given that the nationalization of Italians was a racializing project, the Italian case might be said to reside somewhere in between Arendt's German and English models. Italian racial discourse did not aim to insulate the nobility from the bourgeoisie, and much less the laboring (agricultural) masses. Instead, it sought to articulate (in both senses: to produce discursively and to join together) a hard-working Italian population that was scattered across oceans and seas. It did so with recourse to rallying cries, as in Germany, about freedom from the geographical and physiological fragmentation wrought by foreign occupiers *and* the "hereditary genius" so cherished in the English tradition.

Another central claim throughout the readings of Italian cultural production contained in this book has been that viewing this web through a biopolitical lens allows us a perspective that other analytics, primarily that of anti-racism, do not. The multiple but often invisible or even disavowed intersections between race and (re)productivity that these cultural products—drawn from proto-sociological inquiry, popular hygiene novels, decadent novels, political speeches, verse, and film—harness are revealed when viewed in light of the biopolitical. We might say then that while both Esposito (*Bíos. Biopolitics and Philosophy*) and Hardt and Negri (*Commonwealth*) have sought to turn biopolitics away from its negative, thanatopolitical implications by recasting it in a politically affirmative vein, throughout these pages, we have aimed to activate its analytical productivity. By engaging biopolitics on its own terms—productive, life-affirming, vital—we have been able to uncover how it shaped racial subjectification of the past, and can therefore grasp how it continues to invest Italy's present. While, as we have seen, one primary analytical pitfall of anti-racism in Italy has been its inextricability from anti-fascism, and therefore its relegation of racialist language to a past-tense aberration, biopolitics enables us to see how these enduring concerns about the proper relationship between life and politics belong to our present. Colonial war as life-affirming, state-sponsored infant cremation as immunizing, regulated reproduction as liberating, aerial conquest as therapeutic, bodily sacrifice and mutilation as healing—these paradoxes are just a few key points on the biopolitical constellation that we have been sketching, and they would be profoundly altered if not obscured altogether if our lens were confined to the one requiring us to scrutinize the page for scenes of racial subjection or persecution, or black-and-white logics of inferiority and superiority. And yet the broad and layered field of racial discourse between Italian Unification and World War I to which we have attempted to gain access would appear incomplete, or at the very least much more narrow and superficial, without taking these very paradoxes into account.

As Italy joined the ranks of other modern European democracies (however tardily) at the end of the nineteenth century, and as policymakers, doctors, and artists struggled to define the contours (and hues) of Italian citizenship, they cast their nets far and wide, to Italy's southernmost and north-easternmost regions, across oceans and seas, in a desperate attempt to "capture" all the biological beings they could and ensure their participation (physical, economic, ideological, symbolic) in national life. In so doing, these thinkers forged significant rhetorical bonds between Italian bodies and the lands (regional, national, colonial) they inhabited as they tethered Italian citizenship and national belonging to novel and pre-existing understandings about physiology and physiognomy, somatics and chromatics, blood and soil that we can only describe as racial. What has drawn together the texts under consideration in this book is how they produce racialized Italian subjects in line with biopolitical imperatives. I have mentioned that the extent to which this reached its grisly height in the fascist regime's colonial and racial policies of the 1930s has been well documented. And yet, as I suggested at the beginning of this book, we must be careful not to assume that such imperatives vanished along with the totalitarian regimes that brought them to their most murderous extremes.[18] The preservation of certain forms of life (which contains the seeds of its own opposite: exclusion and/or the negation of other forms of life) remains at the center of the Italian political scene. As long as former Prime Minister Berlusconi strikes historic "friendship" deals with former Libyan dictator Muammar el-Qaddafi to return boatloads

18 Esposito rightly warns us not to allow our necessary condemnation of twentieth-century totalitarianisms (Communism and Nazism) to obscure the specificity of Nazism and to shadow over its persistence in contemporary life. Unlike Communism, which Esposito argues grows out of the ideological and lexical underpinnings of western modernity, Nazism changes the conceptual vocabulary of modernity: "[P]recisely because it lies entirely outside of modern language, because it is situated decidedly *after* it, Nazism embarrassingly brushes up against a dimension that is part of our experience as post-moderns" (Esposito, "Nazism and Us," 80). For Esposito, the specificity of Nazism's language lies in the absolute literalization of the biological metaphor (body-politic, state-body) by political officials, the taking up and eventual overturning of the biopolitical imperative to protect life, so that mass murder was understood as a way of healing the German people ("Jews do not *resemble* parasites, they do not behave *like* bacteria—they *are* such things. And they are treated as such," Esposito, "Nazism and Us," 85). And yet, as Esposito compellingly argues, we have not yet fully emerged from such presuppositions and their effects.

of asylum seekers to Libya (in violation of European Union laws against *refoulement,* or the forced return of migrants to places in which they risk persecution) and, following a curious logic, does so in the name of an apology for Italian colonial atrocities there; as long as Italians vote overwhelmingly to stop the privatization of water and defend it as a *bene comune,* for the common good, as opposed to a market-driven commodity;[19] as long as Catholic church groups organize public forums on bioethics and offer temporary shelter to migrants; and as long as migrant workers in southern Italy protest their abysmal working and living conditions; in short, as long politics draws real and imagined boundaries around life, and as long as qualifying life defines the parameters of the political battlefield, these truly are today, as they have been, vital subjects.

19 See: Hardt and Negri, *Commonwealth.*

Bibliography

Abel, Richard. *French Film Theory and Criticism: A History/Anthology, 1907–1939*. 2 vols. Princeton, NJ: Princeton University Press, 1988.

—, ed. *Silent Film*. London: Athlone Press, 1996.

Agamben, Giorgio. *Homo Sacer: Sovereign Power and Bare Life*. Trans. Daniel Heller-Roazen. Stanford, CA: Stanford University Press, 1998.

—. *Means without End: Notes on Politics*. Minneapolis, MN: University of Minnesota Press, 2000.

—. *State of Exception*. Trans. Kevin Attell. Chicago, IL: University of Chicago Press, 2005.

—. *What is an Apparatus? And Other Essays*. Stanford, CA: Stanford University Press, 2009.

Agnew, John. *Place and Politics in Modern Italy*. Chicago, IL: University of Chicago Press, 2002.

—. "Time into Space. The Myth of 'Backward' Italy in Modern Europe." *Time and Society* 5.1 (1996): 27–45.

Ahmida, Ali Abdullatif. *The Making of Modern Libya: State Formation, Colonization, and Resistance, 1830–1932*. 1994. Albany, NY: State University of New York Press, 2009.

Alatri, Paolo. "Ideologia e politica in D'Annunzio." *D'Annunzio a Yale: atti del convegno*. Gardone Riviera: Fondazione il Vittoriale degli italiani, 1989. 23–35.

Alessandrini, Cecilia. "Gorgona: l'ultima colonia penale agricola d'Italia e d'Europa." *L'Huffington Post (Gruppo Espresso)* June 24, 2014. www.huffingtonpost.it/cecilia-alessandrini/gorgona-colonia-penale-italia_b_5524897.html (accessed July 9, 2014).

Alfieri, Vittorio. *Il misogallo: prose e rime*. Florence: Maza, 1946.

—. *Vita*. Ed. V. Branca. Milan: Mursia, 1983.

Alighieri, Dante. *Divina commedia*. Ed. G. Petrocchi and A. Ciotti. Padua: R.A.D.A.R., 1968.

—. *The Divine Comedy of Dante Alighieri. Purgatorio*. Trans. Robert Durling. Oxford; New York: Oxford University Press, 2003.

Allen, Beverly, and Mary J. Russo. *Revisioning Italy: National Identity and Global Culture*. Minneapolis, MN: University of Minnesota Press, 1997.

Alovisio, Silvio. "Il film che visse due volte. *Cabiria* tra antichi segreti e nuove ricerche." *Cabira & Cabiria*. Ed. Silvio Alovisio and Alberto Barbera. Turin: Museo Nazionale del Cinema, 2006. 15–53.

—. "The 'Pastrone System': Itala Film from the Origins to World War I." *Film History* 12 (2000): 250–261.

Alovisio, Silvio, and Alberto Barbera, eds. *Cabiria & Cabiria*. Turin: Museo Nazionale del Cinema, 2006.

Anderson, Benedict. *Imagined Communities*. London: Verso, 1991.

Andreoli, Annamaria. *D'Annunzio*. Bologna: Mulino, 2004.

—. *Il viviere inimitabile. Vita di Gabriele D'Annunzio*. Milan: Mondadori, 2000.

Arendt, Hannah. *The Origins of Totalitarianism*. 1951. Orlando, FL: Harcourt Brace, 1973.

—. "Race-Thinking before Racism." *The Review of Politics* 6.1 (1944): 36–73.

Argentieri, Mino. *L'occhio del regime. Informazione e propaganda nel cinema del fascismo*. Florence: Vallecchi, 1979.

Arminese, Gabriella. *La pedagogia "igienica" di Paolo Mantegazza*. Lecce: Pensa, 2003.

Arslan, Antonia, and Margherita Ganazzoli. "Neera e Paolo Mantegazza: storia di una collaborazione." *Rassegna della letteratura italiana* 87.1–2 (Winter 1983): 102–124.

Ascoli, Albert Russell, and Krystyna Clara Von Henneberg. *Making and Remaking Italy: The Cultivation of National Identity around the Risorgimento*. Oxford; New York: Berg, 2001.

Ashcroft, Bill, Gareth Griffiths, and Helen Tiffin. *The Postcolonial Studies Reader*. New York; London: Routledge, 1998.

Asor Rosa, Alberto. *La cultura. Storia d'Italia: Dall'Unità a oggi*. Turin: Einaudi, 1975.

Atkinson, David. "Encountering Bare Life in Italian Libya and Colonial Amnesia in Agamben." *Agamben and Colonialism*. Ed. Marcello Svirsky and Simone Bignall. Edinburgh: Edinburgh University Press, 2012. 155–177.

Atkinson, David, and Denis Cosgrove. "Urban Rhetoric and Embodied Identities: City, Nation, and Empire at the Vittorio Emanuele II Monument in Rome, 1870–1945." *Annals of the Association of American Geographers* 88.1 (1998): 28–49.

Atti Parlamentari. Camera dei Deputati, December 1887. c. 318. Vol. Leg XVI.

Bacchelli, Riccardo. *Mal d'Africa. Romanzo storico*. Milan: Garzanti, 1944.

Balibar, Etienne. "The Nation Form: History and Ideology." *Race, Nation, and Class. Ambiguous Identities*. Ed. Etienne Balibar and Immanuel Wallerstein. New York: Verso, 1991. 86–106.

Banti, Alberto. *La nazione del Risorgimento. Parentela, santità e onore alle origini dell'Italia unita*. Turin: Einaudi, 2000.

—. *Il Risorgimento italiano*. Rome: Laterza, 2004.

Baranello, Adriana. "Giovanni Pascoli's 'La grande proletaria si è mossa.' A Translation and Critical Introduction." *California Italian Studies* 2:1 (2011).

Baratieri, Daniela. *Memories and Silences Haunted by Fascism: Italian Colonialism*. Bern: Peter Lang, 2010.

Bàrberi Squarotti, Giorgio. *Invito alla lettura di Gabriele D'Annunzio*. Milan: Mursia, 1982.

Barbujani, Guido. *Questione di razza: romanzo*. Milan: Mondadori, 2003.

Barilli, Renato. *D'Annunzio in prosa*. Milan: Mursia, 1993.

Barrera, Giulia. "Colonial Affairs: Italian Men, Eritrean Women, and the Construction of Racial Hierarchies in Colonial Eritrea (1885–1941)." Ph.D. Dissertation, History. Northwestern University, 2002.

—. "Sex, Citizenship, and the State: The Construction of Public and Private Spheres in Colonial Eritrea." *Gender, Family, and Sexuality. The Private Sphere in Italy, 1860–1945*. Ed. Perry Wilson. New York: Palgrave, 2004. 157–172.

Barthes, Roland. *Camera Lucida. Reflections on Photography*. London: Vintage, 2000.

Bartolommei Gioli, Gino. "Il programma dell'Agricoltura Coloniale." *L'Agricoltura coloniale* 1.1 (July 1907): 1–5.

Baxter, Peter. "On the History and Ideology of Film Lighting." *Screen* 16.3 (Autumn 1975): 83–106.

Bazzicalupo, Laura. "The Ambivalences of Biopolitics." *Diacritics* 36.2 (Summer 2006): 109–116.

—. *Biopolitica: Una mappa concettuale*. Rome: Carocci, 2010.

Becker, Jared. "D'Annunzio, Orientalism and Imperialism." *Stanford Italian Review* IX.1–2 (1990): 87–103.

—. *Nationalism and Culture: Gabriele D'Annunzio and Italy after the Risorgimento*. New York: P. Lang, 1994.

Behdad, Ali. *Belated Travelers: Orientalism in the Age of Colonial Dissolution*. Durham, NC; London: Duke University Press, 1994.

Ben-Ghiat, Ruth. "Envisioning Modernity: Desire and Discipline in the Italian Fascist Film." *Critical Inquiry* 23 (Autumn 1996): 109–144.

—. *Fascist Modernities: Italy, 1922–1945*. Berkeley, CA: University of California Press, 2001.

—. "The Italian Colonial Cinema: Agendas and Audiences." *Italian Colonialism*. Ed. Ruth Ben-Ghiat and Mia Fuller. New York: Palgrave, 2005. 179–192.

—. *Italian Fascism's Empire Cinema*. Bloomington: Indiana University Press, 2015.

Ben-Ghiat, Ruth, and Mia Fuller, eds. *Italian Colonialism*. New York: Palgrave Macmillan, 2005.

Benjamin, Walter. *The Work of Art in the Age of its Technological Reproducibility, and Other Writings on Media*. Ed. Michael William Jennings, Brigid Doherty, et al. Cambridge, MA: Belknap Press of Harvard University Press, 2008.

—. *The Arcades Project*. Cambridge, MA: Belknap Press of Harvard University Press, 2002.

Berardi, Franco "Bifo". *After the Future*. Oakland, CA: AK Press, 2011.

Beretto, Paolo, and Gianni Rondolino, eds. *Cabiria e il suo tempo*. Turin: Museo Nazionale del Cinema, 1998.

Bernardini, Aldo. *Cinema muto italiano*. 3 vols. Rome: Laterza, 1980.

Bertazzoli, Raffaella. "I messaggi 'Agli Italiani degli Stati Uniti' tra interventismo e impresa fiumana." *D'Annunzio a Yale: atti del convegno*. Ed. Paolo Valesio. Gardone Riviera: Fondazione il Vittoriale degli italiani, 1989. 347–361.

Bertellini, Giorgio. "Colonial Autism. Whitened Heroes, Auditory Rhetoric and National Identity in Interwar Italian Cinema." *A Place in the Sun. Africa in Italian Colonial Culture from Post-Unification to the Present*. Ed. Patrizia Palumbo. Berkeley; Los Angeles, CA: University of California Press, 2003. 255–278.

—. "Introduction: Early Italian Cinema." *Film History* 12 (2000): 235–239.

—. *Italy in Early American Cinema. Race, Landscape and the Picturesque*. Bloomington, IN: Indiana University Press, 2010.

—. "Southern Crossings: Italians, Cinema and Modernity (Italy, 1861–New York, 1920)." Ph.D. Dissertation, Cinema Studies. New York University, 2001.

Bhabha, Homi K. *The Location of Culture*. London; New York: Routledge, 1994.

Bidussa, David. *Il mito del bravo italiano*. Milan: Saggiatore, 1994.

Bishop, Ryan, and John Phillips. *Modernist Avant-Garde Aesthetics and Contemporary Military Technology: Technicities of Perception*. Edinburgh: Edinburgh University Press, 2010.

Bocca, Giorgio. *La disunità degli italiani*. Milan: Garzanti, 1990.

Boggio, Cecilia. "Black Shirts/Black Skins. Fascist Italy's Colonial Anxieties and *Lo squadrone bianco*." *A Place in the Sun. Africa in Italian Colonial Culture from Post-Unification to the Present*. Ed. Patrizia Palumbo. Berkeley; Los Angeles, CA: University of California Press, 2003. 279–298.

Bollati, Giulio. *L'italiano. Il carattere nazionale come storia e come invenzione*. Turin: Einaudi, 1983.

Bonavita, Riccardo, et al. *L'offesa della razza: razzismo e antisemitismo dell'Italia fascista*. Emilia Romagna, biblioteche archivi. Bologna: Pàtron, 2005.

Bonetta, Gaetano. *Corpo e nazione. L'educazione ginnastica, igienica e sessuale nell'Italia liberale*. Milan: FrancoAngeli, 1990.

Bongie, Chris. *Exotic Memories: Literature, Colonialism and the Fin de Siècle*. Stanford, CA: Stanford University Press, 1991.

Boni, Monica. *L'erotico senatore. Vita e studi di Paolo Mantegazza*. Genoa: Name, 2002.

Bonomi, Ivanoe. *Leopoldo Franchetti e il Mezzogiorno*. Rome: Tipografia Ed. Italia, 1918.

Brondino, Michele. *La stampa italiana in Tunisia. Storia e società. 1838–1956*. Milan: Jaca Book, 1998.

Brooks, Peter. *Body Work: Objects of Desire in Modern Narrative*. Cambridge, MA.: Harvard University Press, 1993.

Brunetta, Gian Piero. *Cinema italiano tra le due guerre. Fascismo e politica cinematografica*. Milan: Mursia, 1975.

—. *The History of Italian Cinema: A Guide to Italian Film from its Origins to the Twenty-First Century*. Trans. Jeremy Parzen. Princeton, NJ: Princeton University Press, 2009.

Brunetta, Gian Piero, and Jean Gili. *L'ora d'Africa del cinema italiano (1911–1989)*. Rovereto: Materiali di lavoro, 1990.

Bruno, Giuliana. *Streetwalking on a Ruined Map. Cultural Theory and the City Films of Elvira Notari*. Princeton, NJ: Princeton University Press, 1993.

Bucci, Sante. "La scuola della Montesca. Un centro educativo internazionale." *Leopoldo e Alice Franchetti e il loro tempo*. Ed. Paolo Pezzino and Alvaro Tacchini. Città di Castello: Petruzzi, 2002: 195–242.

Burgio, Alberto, ed. *Nel nome della razza: il razzismo nella storia d'Italia 1870–1945*. Bologna: Il Mulino, 1999.

Burgio, Alberto, and Luciano Casali, eds. *Studi sul razzismo italiano*. Bologna: CLUEB, 1996.

Burgwyn, H. James. *The Legend of the Mutilated Victory: Italy, the Great War, and the Paris Peace Conference, 1915–1919*. Westport, CT: Greenwood Press, 1993.

Butters, Gerard. *Black Manhood on the Silent Screen*. Lawrence, KS: University of Kansas Press, 2002.

Campanella, Tommaso. *La città del sole*. 1602. Trans. Francesco Bartoletta. Cosenza: Edizioni Brenner, 1985.

Campbell, Timothy. "Bios, Immunity, Life: The Thought of Roberto Esposito." Esposito, Roberto. *Bíos. Biopolitics and Philosophy*. Minneapolis, MN: University of Minnesota Press, 2008. vii–xliii.

—. "Interview with Roberto Esposito." *Diacritics* 36.2 (Summer 2006): 49–56.

—. "'Infinite Remoteness': Marinetti, Bontempelli, and the Birth of Modern Italian Visual Culture." *MLN* 120 (January 2005): 111–136.

—. *Wireless Writing in the Age of Marconi*. Minneapolis, MN: University of Minnesota Press, 2006.

Canjels, Rudmer. *Distributing Silent Film Serials: Local Practices, Changing Forms, Cultural Transformation*. New York: Routledge, 2010.

Canosa, Michele, ed. *1905. La presa di Roma: alle origini del cinema italiano*. Bologna; Genova: Cineteca Bologna; Le mani, 2006.

—. *Cinema muto italiano: tecnica e tecnologia*. 2 vols. Rome: Carocci, 2006.

Canudo, Ricciotto. *L'usine aux images*. Geneva: Office Central d'Édition, 1927.
Capatti, Alberto. "Preface." *Paolo Mantegazza. L'Anno 3000. Sogno*. Bergamo: Lubrina, 1988. 9–19.
Caprotti, Federico. *Mussolini's Cities: Internal Colonialism in Italy, 1930–1939*. Youngstown, OH: Cambria Press, 2007.
—. "Scipio Africanus: Film, Internal Colonization, and Empire." *Cultural Geographies* 16.3 (2009): 381–401.
Caraccio, Armand. *D'Annunzio, dramaturge*. Paris: Presses universitaires de France, 1951.
Carpi, Leone. *Delle colonie e dell'emigrazione d'Italiani all'estero*. Volume 3. Milan: Lombarda, 1874.
Cartwright, Lisa, and Brian Goldfarb. "On the Subject of Neural and Sensory Prosthesis." *The Prosthetic Impulse: From a Posthuman Present to a Biocultural Future*. Ed. M. Smith and J. Morra. Cambridge, MA: MIT Press, 2006. 125–154.
Cassata, Francesco. *Molti, sani e forti. L'eugenetica in Italia*. Turin: Bollati Boringhieri, 2006.
—. "Rigenerare la razza. La via italiana all'eugenica tra Lombroso e Pareto." *Storia della Shoah in Italia*. Ed. M.A. Matard-Bonucci et al. Turin: UTET, 2010. 115–138.
Cavarero, Adriana. "Diotima." *Nonostante Platone. Figure femminili nella filosofia antica*. Rome: Riuniti, 1999. 93–122.
—. *Stately Bodies: Literature, Philosophy, and the Question of Gender*. Trans. Deanna Shemek. Ann Arbor, MI: University of Michigan Press, 2002.
Chabod, Federico. *L'idea di nazione*. Rome; Bari: Laterza, 1974.
Chakrabarty, Dipesh. *Provincializing Europe: Postcolonial Thought and Historical Difference*. Princeton, NJ: Princeton University Press, 2000.
Chambers, Ross. "The Unexamined." *Whiteness: A Critical Reader*. Ed. Mike Hill. New York: New York University Press, 1997. 187–203.
Chatterjee, Partha. *Nationalist Thought and the Colonial World: A Derivative Discourse*. Minneapolis, MN: University of Minnesota Press, 1993.
Cherchi Usai, Paolo. "*Cabiria*, an Incomplete Masterpiece: The Quest for the Original 1914 Version." *Film History* 2 (1988): 155–166.
—. *Giovanni Pastrone: gli anni d'oro del cinema a Torino*. Turin: UTET, 1986.
—. *Silent Cinema. An Introduction*. London: British Film Institute, 2000.
Chiarelli, Cosimo. "Mantegazza e la fotografia: un'antologia di immagini." *Paolo Mantegazza. Medico, antropologo, viaggiatore*. Ed. Cosimo Chiarelli and Walter Pasini. Florence: Firenze University Press, 2002. 93–115.
Chibber, Deepak. *Postcolonial Theory and the Specter of Capital*. New York: Verso, 2013.
Chimirri, Giovanna Finocchiaro. *D'Annunzio e il cinema.* Cabiria. Catania: Cooperativa Universitaria Editrice Catanese di Magistero, 1986.

Chiozzi, Paolo. "Esistono gli 'Ariani'? Perplessità e contraddizioni di PM in tema di 'Razze'." *Paolo Mantegazza. Medico, antropologo, viaggiatore*. Ed. Cosimo Chiarelli and Walter Pasini. Florence: Firenze University Press, 2002. 43–52.
—. "Fotografia e antropologia nell'opera di Paolo Mantegazza." *Semestrale dell'Archivio fotografico toscano* III.6 (1987): 56–61.
Choate, Mark. *Emigrant Nation: The Making of Italy Abroad*. Cambridge, MA: Harvard University Press, 2008.
—. "From Territorial to Ethnographic Colonies and Back Again: The Politics of Italian Expansion, 1890–1912." *Modern Italy* 8.1 (2003): 65–75.
Clò, Clarissa, and Teresa Fiore. "Unlikely Connections: Italy's Cultural Formations between Home and the Diaspora." *Diaspora: A Journal of Transnational Studies* 10.3 (2001): 415–441.
Colliva, Guiliano. *Uomini e aerei nella storia dell'aeronautica militare italiana*. Milan: Bramante, 1973.
Collotti, Enzo. *Razza e fascismo: la persecuzione contro gli ebrei in Toscana: 1938–1943*. Rome: Carocci, 1999.
Contini, Giovanni. *La memoria divisa*. Milan: Rizzoli, 1997.
Corradini, Enrico. *La conquista di Tripoli*. Milan: Treves, 1912.
—. *L'Ora di Tripoli*. Milan: Treves, 1911.
—. *Sopra le vie del nuovo impero*. Milan: Treves, 1912.
Corriere della sera. "Maciste in persona al *Lirico* nelle rappresentazioni di *Cabiria*." May 9, 1914.
Corti, Maria. "Preface." Paolo Mantegazza and Neera. *Dizionario d'igiene per le famiglie*. Milan: Scheiwiller, 1985.
Cosgrove, Denis. *Social Formation and Symbolic Landscape*. London: Croom Helm, 1984.
Cottini, Luca. "La novità di *Maciste alpino*." *Italian Culture* 71.1 (March 2009): 43–59.
Crary, Jonathan. *Technologies of the Observer*. Cambridge, MA: Massachusetts Institute of Technology, 1990.
Croce, Benedetto. "Scienziati-letterati." *La letteratura della nuova Italia. Saggi critici*. Vol. VI. Rome; Bari: Laterza, 1940.
—. *Storia d'Italia dal 1871 al 1915*. 10th ed. Bari: G. Laterza, 1953.
Cuomo, Franco. *I dieci: chi erano gli scienziati italiani che firmarono il Manifesto della razza*. Milan: Baldini Castoldi Dalai, 2005.
Curcio, Anna, and Miguel Mellino. "La razza al lavoro. Rileggere il razzismo, ripensare l'antirazzismo in Italia." *La razza al lavoro*. Ed. Anna Curcio and Miguel Mellino. Rome: Manifestolibri, 2012: 7–36.
D'Annunzio, Gabriele. *L'Innocente (The Victim)*. Trans. Georgina Harding. Sawtry, UK: Dedalus, 1991.

—. *Merope. Versi d'amore e di gloria*. Ed. Annamaria Andreoli and Niva Lorenzini. Vol. II. 1911–1912. Milan: Mondadori, 1984. 646–757.
—. *La penultima ventura. Scritti e discorsi fiumani*. Ed. R. De Felice. Milan: Mondadori, 1974.
—. "Più che l'amore." *Tragedie, sogni, e misteri*. Vol. 1. Milan: Mondadori, 1968.
—. *Prosa di ricera, di lotta, di comando, di conquista, di tormento, d'indovinamento, di rinnovamento, di celebrazione, di rivendicazione, di liberazione, di favole, di giochi, di baleni*. Milan: Mondadori, 1947.
—. *Prose di ricerca*. Ed. Annamaria Andreoli and Giorgio Zanetti. Vol. I. Milan: Mondadori, 2005.
—. *Prose di ricerca*. Vol. III. Milan: Mondadori, 2005.
—. *Prose di romanzi 1*. Eds. Annamaria Andreoli and Ezio Raimondi. Milan: Mondadori, 1988.
—. *Prose di romanzi 2*. Ed. Niva Lorenzini and Ezio Raimondi. Milan: Mondadori, 1989.
—. *Taccuini*. Milan: Mondadori, 1965.
—. *The Child of Pleasure*. Trans. Georgina Harding. New York: Howard Fertig, 1990.
—. *The Maidens of the Rocks*. Trans. Annetta Halladay-Antona and Giuseppe Antona. Boston, MA: L.C. Page, 1898.
—. *The Triumph of Death*. Trans. Georgina Harding. Sawtry, UK: Dedalus, 1990.
—. *Tutte le novelle*. Meridiani. Ed. Annamaria Andreoli and Marina De Marco. Milan: Mondadori, 1992.
—. *Versi d'amore e di gloria, Vol. II*. Meridiani. Ed. Luciano Anceschi, Annamaria Andreoli and Niva Lorenzini. 3rd. ed. Milan: Mondadori, 1982.
D' Azeglio, Massimo. *I miei ricordi*. Milan: Feltrinelli, 1963.
D' Azeglio, Massimo, and Alberto Maria Ghisalberti. *I miei ricordi*. Florence: Barbera, 1891.
D'Onofrio, Andrea. *Razza, sangue e suolo: utopie della razza e progetti eugenetici nel ruralismo nazista*. Naples: ClioPress, 2007.
Daggett, R.M. and Goodman, J.T. *The Psychoscope. A Sensational Drama in Five Acts*. Ed. Lawrence Berkove. 1871. Charleston, SC: Mark Twain Journal, 2006.
Dagrada, Elena, et al., eds. *Moltiplicare l'istante: Beltrami, Comerio e Pacchioni tra fotografia e cinema*. Milan: Il Castoro, 2007.
Dal Lago, Alessandro. *Non-persone: l'esclusione dei migranti in una società globale*. Milan: G. Feltrinelli, 1999.
Davis, John A. *Italy in the Nineteenth Century 1796–1900*. Oxford; New York: Oxford University Press, 2002.
—. "The South, the Risorgimento, and the Origins of the 'Southern Problem.'" *Gramsci and Italy's Passive Revolution*. Ed. John Davis. New York: Barnes and Noble, 1979. 67–103.

De Amicis, Edmondo. *Cuore*. Milan: Mondadori, 1984.

—. "Dagli Apennini alle Ande." *Opere scelte*. Ed. Folco Portinari and Giusi Baldissone. Milan: Mondadori, 1995. 301–333.

—. *Roma capitale*. Ed. Ugo Piscopo. Naples: Alfredo Guida Editore, 1995.

De Bernardi, A., and L. Ganapini, eds. *Storia d'Italia 1860–1995*. Milan: Mondadori, 1996.

De Felice, Renzo. *D'Annunzio politico, 1918–1938*. Rome; Bari: Laterza, 1978.

De Franceschi, Leonardo. *L'Africa in Italia: per una controstoria postcoloniale del cinema italiano*. Rome: Aracne, 2013.

De Grazia, Victoria. *The Culture of Consent: Mass Organization of Leisure in Fascist Italy*. Cambridge: Cambridge University Press, 1981.

De Oliveira, João S. "*Cabiria*, una nuova sfida per il restauro." *Cabiria & Cabiria*. Ed. Silvio Alovisio and Alberto Barbera. Turin: Museo Nazionale del Cinema, 2006. 54–61.

Dean, Jodi. "Drive as the Structure of Biopolitics. Economy, Sovereignty, and Capture." *Krisis. Journal for Contemporary Philosophy* 2 (2010): 1–15.

—. *Žižek's Politics*. New York: Routledge, 2006.

Dean, Mitchell. *Governmentality. Power and Rule in Modern Society*. London: Sage, 1999.

Debord, Guy. *Society of the Spectacle*. Detroit: Black and Red, 1983.

Del Boca, Angelo. *Gli italiani in Africa Orientale, Vol. 1*. Rome: Laterza, 1976.

—. "Realtà e leggenda di Dogali." *Gli italiani in Africa Orientale. Vol. 1*. Rome: Laterza, 1976. 239–276.

Deleuze, Gilles. "What is a *Dispositif*?" *Michel Foucault: Philosopher*. Ed. Timothy Armstrong. New York: Routledge, 1991. 159–168.

Deleuze, Gilles, and Felix Guattari. "What is a Minor Literature?" *Kafka: Toward a Minor Literature*. Minneapolis, MN: University of Minnesota Press, 1986. 16–27.

Demetz, Peter. "Aviators at Brescia: Selective Views." *The Airshow at Brescia, 1909*. New York: Farrar, Straus and Giroux, 2002. 186–223.

—. "D'Annunzio: Poet and Aviator." *The Airshow at Brescia, 1909*. New York: Farrar, Straus and Giroux, 2002. 149–185.

Dickie, John. *Cosa Nostra: A History of the Sicilian Mafia*. Basingstoke, UK: Palgrave Macmillan, 2004.

—. *Darkest Italy: The Nation and Stereotypes of the Mezzogiorno, 1860–1900*. New York: St. Martin's Press, 1999.

Duncan, Derek. "Italy's Postcolonial Cinema and its Histories of Representation." *Italian Studies* 63.2 (2008): 195–211.

—. *Reading and Writing Italian Homosexuality: A Case of Possible Difference*. Hampshire, England; Burlington, VT: Ashgate, 2006.

Duncan, Derek and Jacqueline Andall. *Italian Colonialism: Legacy and Memory*. Oxford; Bern: Peter Lang, 2005.

—, eds. *National Belongings. Hybridity in Italian Colonial and Postcolonial Cultures.* Oxford; Bern: Peter Lang, 2010.

Dyer, Richard. *White.* London; New York: Routledge, 1997.

Esposito, Roberto. *Bios: Biopolitica e filosofia.* Turin: Einaudi, 2004.

—. *Bíos. Biopolitics and Philosophy.* Trans. Timothy Campbell. Minneapolis, MN: University of Minnesota Press, 2008.

—. *Communitas: origine e destino della comunità.* Turin: Einaudi, 1998.

—. *Communitas: The Origin and Destiny of Community.* Trans. Timothy Campbell. Stanford, CA: Stanford University Press, 2009.

—. *Immunitas.* Trans. Zakiya Hanafi. Malden, MA: Polity Press, 2012.

—. *Immunitas: protezione e negazione della vita.* Turin: Einaudi, 2002.

—. "Nazism and Us." *Terms of the Political: Community, Immunity, Biopolitics.* New York: Fordham University Press, 2013. 79–88.

—. *Termini della politica. Comunità, immunità, biopolitica.* Milan: Mimesis, 2008.

—. *Terms of the Political: Community, Immunity, Biopolitics.* Trans. Rhiannon Noel Welch. New York: Fordham University Press, 2013.

—. *Terza persona. Politica della vita e filosofia dell'impersonale.* Turin: Einaudi, 2007.

Essed, Philomena, and David Theo Goldberg, eds. *Race Critical Theories: Text and Context.* Malden, MA.: Blackwell Publishers, 2002.

Evangelisti, Giorgio. *La scrittura nel vento.* Florence: Olimpia, 1999.

Evola, Julius, and Piero Di Vona. *I testi de La difesa della razza.* Padua: Edizioni di Ar, 2001.

Evola, Julius, and G.F. Lami. *La nobiltà della stirpe : 1932–1938; La difesa della razza : 1939–1942.* Biblioteca evoliana. Roma: Fondazione Julius Evola, 2002.

Fabian, Johannes. *Time and the Other: How Anthropology Makes its Object.* New York: Columbia University Press, 2002.

Faccioli, Alessandro. "Pastrone, Camasio, D'Annunzio, e *Cabiria*: trent'anni dopo." *Cabiria & Cabiria.* Ed. Silvio Alovisio and Alberto Barbera. Turin: Museo Nazionale del Cinema, 2006. 335–348.

Falasca-Zamponi, Simonetta. *Fascist Spectacle: The Aesthetics of Power in Mussolini's Italy.* Studies on the History of Society and Culture 28. Berkeley, CA: University of California Press, 1997.

Fanon, Frantz. *Black Skin, White Masks.* Trans. Richard Philcox. New York: Grove Press, 2008.

Farassino, Alberto. "Anatomia del cinema muscolare." *Gli uomini forti.* Ed. Alberto Farassino and Tatti Sanguinetti. Milan: Mazzotta, 1983. 29–51.

—. "Maciste e il paradigma divistico." *Cabiria e il suo tempo.* Ed. Paolo Beretto and Gianni Rondolino. Turin: Museo Nazionale del Cinema, 1998. 223–250.

Fern, Chris. *Narrating Utopia: Ideology, Gender, Form in Utopian Literature.* Liverpool: Liverpool University Press, 1999.

Ferrari, Chiara. *The Rhetoric of Violence and Sacrifice in Fascist Italy: Mussolini, Gadda, Vittorini*. Toronto: University of Toronto Press, 2013.

Ferroni, Giulio, ed. *Storia della letteratura italiana. Dall'Ottocento al Novecento*. Milan: Einaudi, 1991.

Filippucci, Paola. "Anthropological Perspectives on Culture in Italy." *Italian Cultural Studies: An Introduction*. Ed. D. Forgacs and R. Lumley. New York; Oxford: Oxford University Press, 1996.

Finaldi, Giuseppe Maria. "Defining a 'Culture of Colonialism' and the Italian Case." *Italian National Identity and The Scramble for Africa. Italy's African Wars in the Era of Nation-building, 1870–1900*. Bern: Peter Lang, 2009. 33–52.

Fiore, Teresa. "The Emigrant Post-'Colonia' in Contemporary Italy." *Postcolonial Italy: Challenging National Homogeneity*. Ed. Cristina Lombardi-Diop and Caterina Romeo. New York: Palgrave, 2012. 71–82.

Flaxman, Gregory. "Introduction." *The Brain is the Screen: Deleuze and the Philosophy of Cinema*. Ed. Gregory Flaxman. Minneapolis, MN: University of Minnesota Press, 2000. 1–60.

Flynn, Thomas. "Foucault and the Eclipse of Vision." *Modernity and the Hegemony of Vision*. Ed. David Michael Levin. Berkeley; Los Angeles, CA: University of California Press, 1993. 273–286.

Fogu, Claudio. "Fascism and Historic Representation: The 1932 Garibaldian Celebrations." *Journal of Contemporary History* 31.2 (1996): 317–345.

Foot, John. *Italy's Divided Memory*. New York: Palgrave, 2009.

Forgacs, David. "Colonies." *Italy's Margins. Social Exclusion and Nation Formation since 1861*. Cambridge, New York: Cambridge University Press, 2014. 67–138.

Forti, Simona. "The Biopolitics of Souls: Racism, Nazism, and Plato." *Political Theory* 34.1 (February 2006): 9–32.

Fortunato, Giustino. *Leopoldo Franchetti*. Ricordi. Rome, Laziale 1918.

Foscolo, Ugo. *Le ultime lettere di Jacopo Ortis*. Milan: Rizzoli, 1949.

—. "The Last Letters of Jacopo Ortis." *Three Italian Epistolary Novels: Foscolo, De Meis, Piovene*. Trans. Vincenzo Traversa. New York: Peter Lang, 2005.

Fossati, Roberta. "Il lavoro culturale e la vita affettiva di Alice Hallgarten Franchetti." *Leopoldo e Alice Franchetti e il loro tempo*. Ed. Paolo Pezzino and Alvaro Tacchini. Città di Castello: Petruzzi, 2002. 158–194.

Foucault, Michel. *The Birth of Biopolitics. Lectures at the Collège de France, 1978–1979*. New York: Palgrave, 2008.

—. *Discipline and Punish: The Birth of the Prison*. Trans. Alan Sheridan. New York: Random House, 1977.

—. *"Il faut défendre la société": cours au Collège de France (1975–1976)*. Paris: Seuil/Gallimard, 1997.

—. "Governmentality." *The Essential Foucault: Selections from the Essential Works of Foucault. 1954–1974*. Ed. Paul Rabinow and Nikolas Rose. New York: New Press, 2003. 229–245.

—. *The History of Sexuality, Volume 1.* New York: Vintage Books, 1990.
—. *The Order of Things.* New York: Random House, 1970.
—. *Power/Knowledge.* Ed. Colin Gordon. New York: Pantheon, 1980.
—. *Security, Territory, Population. Lectures at the Collège de France, 1977–78.* New York: Palgrave, 2007.
—. *Society Must Be Defended: Lectures at the Collège de France, 1975–1976.* New York: Picador, 2003.
—. "The Subject and Power." *The Essential Foucault: Selections from the Essential Works of Foucault 1954–1974.* Ed. Paul Rabinow and Nikolas Rose. New York: New Press, 2003. 126–144.
Franchetti, Leopoldo. *Mezzogiorno e colonie.* Ed. Umberto Zanotti-Bianco. Florence: La nuova Italia, 1950.
—. "Prefazione." *L'Africa dopo la Guerra e l'Italia.* Ed. Orazio Pedrazzi. Florence: Stabilimento Pellas, 1917. iii–x.
Franchetti, Leopoldo, and Sidney Sonnino. *Inchiesta in Sicilia.* Florence: Vallecchi, 1974.
Franco, Giuseppe. *Le gemelle africane, ossia l'Africa interna descritta dal vero.* Prato: Tipografia Giachetti, 1882.
Frank, Jason. *Constituent Moments: Enacting the People in Postrevolutionary America.* Durham, NC: Duke University Press, 2010.
Frankenberg, Ruth. "White Women, Race Matters." *Theories of Race and Racism: A Reader.* Ed. Les Back and John Solomos. London and New York: Routledge, 2006. 519–533.
Freud, Sigmund. *Dora. An Analysis of a Case of Hysteria.* Ed. Philip Rieff. New York: Collier, 1963.
—. *Jokes and their Relation to the Unconscious.* The Standard Edition of the Complete Psychological Works of Sigmund Freud. Ed. James Strachey. Vol. VIII. London: Hogarth Press and the Institute of Psychoanalysis, 1978.
Fuller, Mia. "Libyan Genocide 2.0: Today's Conflict in Light of a Colonial Past." *Fair Observer.* September 6, 2011.
—. *Moderns Abroad: Architecture, Cities and Italian Imperialism.* London; New York: Routledge, 2007.
—. "Tradition as a Means to the End of Tradition: Farmers' Houses in Italy's Fascist-Era New Towns." *The End of Tradition?* Ed. Nezar AlSayyad. London; New York: Routledge, 2004. 171–186.
—. "Wherever You Go, There You Are: Fascist Plans for the Colonial City of Addis Ababa and the Colonizing Suburb of EUR '42." *Journal of Contemporary History* 21.2 (1996): 397–418.
Gabaccia, Donna R. *Italy's Many Diasporas.* Seattle, WA: University of Washington Press, 2000.
Galli della Loggia, Ernesto. *L'identità italiana.* Bologna: Mulino, 1998.

Galton, Francis. "Hereditary Talent and Character." *Macmillan's Magazine* 12 (1865): 157–166; 318–327.
—. *Inquires into Human Faculty and its Development*. London: Macmillan, 1883.
Gambarota, Paola. *Irresistible Signs: The Genius of Language and Italian National Identity*. Toronto: University of Toronto Press, 2011.
Gates, Henry Louis. *"Race," Writing, and Difference*. Chicago, IL: University of Chicago Press, 1986.
Gentile, Emilio. *Il culto del littorio: la sacralizzazione della politica nell'Italia fascista*. Rome; Bari: Editori Laterza, 1993.
—. "Fascism as Political Religion." *Journal of Contemporary History* 25 (1990): 229–251.
Germinario, Francesco. *Razza del sangue, razza dello spirito: Julius Evola, l'antisemitismo e il nazionalsocialismo: 1930–43*. Turin: Bollati Boringhieri, 2001.
Ghirardo, Diane. *Building New Communities: New Deal America and Fascist Italy*. Princeton, NJ: Princeton University Press, 1989.
Gibson, Mary. "Labeling Woman Deviant: Heterosexual Women, Prostitutes, and Lesbians in Early Criminology Criminological Discourse." *Gender, Family and Sexuality. The Private Sphere in Italy, 1860–1945*. Ed. Perry Wilson. New York: Palgrave, 2004. 89–104.
Gillette, Aaron. *Racial Theories in Fascist Italy*. London; New York: Routledge, 2002.
Gilroy, Paul. *Against Race: Imagining Political Culture Beyond the Color Line*. Cambridge, MA: Belknap Press of Harvard University Press, 2000.
—. "The End of Anti-Racism." *Race Critical Theories: Text and Context*. Ed. David Theo Goldberg and Philomena Essed. Malden, MA: Blackwell, 2002. 249–264.
Gioberti, Vincenzo. *Del primato morale e civile degli italiani*. Naples: Morano, 1866.
Giordano, Cristiana. *Migrants in Translation. Caring and the Logics of Difference in Contemporary Italy*. Berkeley: University of California Press, 2014.
Girolamo, S. "L'emigrazione e le colonie." *Nuova antologia* 27 (Nov 1, 1874): 621–650.
Giuliani, Gaia, and Cristina Lombardi-Diop. *Bianco e nero. Storia dell'identità razziale degli italiani*. Florence: Le Monnier-Mondadori, 2013.
Gnisci, Armando. *Creolizzare l'Europa: letteratura e migrazione*. Rome: Meltemi, 2003.
Goldberg, David Theo. *The Racial State*. Malden, MA; Oxford: Blackwell, 2002.
—. *Racist Culture: Philosophy and the Politics of Meaning*. Oxford; Cambridge, MA: Blackwell, 1993.
Gordon, Colin. "Governmental Rationality: An Introduction." *The Foucault Effect. Studies in Governmentality*. Chicago, IL: Chicago University Press, 1991. 1–52.

Goudet, Jacques. *D'Annunzio romanziere*. Florence: Olschki, 1976.
Gramsci, Antonio. *Quaderni dal carcere*. Turin: Einaudi, 2001.
—. *La questione meridionale*. Ed. F. De Felice and V. Parlato. Rome: Riuniti, 1974.
—. *Selections from the Prison Notebooks*. 1971. Ed. Q. Hoare and G. Smith. New York: International, 2003.
—. *The Southern Question*. Ed. Pasquale Verdicchio. West Lafayette, IN: Bordighera, 1995.
Greene, Shelleen. *Equivocal Subjects. Between Italy and Africa: Constructions of Racial and National Identity in the Italian Cinema*. London; New York: Continuum Press, 2012.
Gribaudi, Gabriella. "Images of the South." *Italian Cultural Studies: An Introduction*. Ed. D. Forgacs and R. Lumley. New York; Oxford: Oxford University Press, 1996. 72–86.
Grieveson, Lee, and Peter Kramer, eds. *The Silent Cinema Reader*. New York: Routledge, 2004.
Guerin, Frances. *A Culture of Light: Cinema and Technology in 1920s Germany*. Minneapolis, MN: University of Minnesota Press, 2005.
Guglielmo, Jennifer, and Salvatore Salerno, eds. *Are Italians White? How Race is Made in America*. New York: Routledge, 2003.
Guha, Ranajit, and Gayatri Spivak, eds. *Selected Subaltern Studies*. New York; Oxford: Oxford University Press, 1988.
Gumbrecht, Hans Ulrich. "I redentori della vittoria. On Fiume's Place in the Genealogy of Fascism." *Journal of Contemporary History* 31.2 (1996): 253–272.
Gunning, Tom. "A Little Light on a Dark Subject." *Critical Quarterly* 45.4 (2003): 50–69.
Gutmann, Amy, ed. *Multiculturalism: Examining the Politics of Recognition*. Princeton, NJ: Princeton University Press, 1994.
Hall, Stuart. "Race, Articulation and Societies Structured in Dominance." *Race Critical Theories: Text and Context*. Ed. David Theo Goldberg and Philomena Essed. Malden, MA: Blackwell, 2002. 38–68.
Hannaford, Ivan. *Race: The History of an Idea in the West*. Washington, D.C.: Woodrow Wilson Center Press, 1996.
Hardt, Michael, and Antonio Negri. *Commonwealth*. Cambridge, MA: Harvard University Press, 2009.
—. *Empire*. Cambridge, MA: Harvard University Press, 2000.
—. *Multitude. War and Democracy in the Age of Empire*. New York: Penguin Press, 2004.
Harper, Graeme, and Jonathan Rayner, eds. *Cinema and Landscape*. Chicago, IL: Intellect, The University of Chicago Press, 2010.
Harrowitz, Nancy. *Antisemitism, Misogyny, and the Logic of Cultural Difference: Cesare Lombroso and Matilde Serao*. Lincoln, NE: University of Nebraska Press, 1994.

Harvey, David. *A Companion to Marx's* Capital, Vol. 1. London, New York: Verso, 2010.

Hearder, Harry. *Italy in the Age of the Risorgimento 1790–1870*. London: Longman, 1983.

Hearn, Jeff. "Gender: Biology, Nature, and Capitalism." *Cambridge Companion to Marx*. Ed. Terrell Carver. Cambridge: Cambridge University Press, 1991. 222–245.

Hettner, Alfred, ed. *Geographische Zeitschrift*. Leipzig: Druck und Verlag, 1903.

Himmelweit, Susan. "Reproduction and the Materialist Conception of History: A Feminist Critique." *Cambridge Companion to Marx*. Ed. Terrell Carver. Cambridge: Cambridge University Press, 1991. 196–221.

Holden, Philip, and Richard Ruppel, eds. *Imperial Desire: Dissident Sexualities and Colonial Literature*. Minneapolis, MN: University of Minnesota Press, 2003.

Hom, Stephanie Malia. "On the Origins of Making Italy: Massimo D'Azeglio and 'Fatta l'Italia, bisogna fare gli Italiani'." *Italian Culture* 31 (March 2013): 1–16.

Horn, David G. *The Criminal Body: Lombroso and the Anatomy of Deviance*. New York: Routledge, 2003.

—. *Social Bodies: Science, Reproduction, and Italian Modernity*. Princeton, NJ: Princeton University Press, 1994.

Hunt, Alastair, and Matthias Rudolf. "Introduction. The Romantic Rhetoric of Life." *Romantic Circles* (Special Issue: "Romanticism and Biopolitics") December 2012. www.rc.umd.edu/praxis/biopolitics/index.html (accessed June 16, 2015).

Inda, Jonathan Xavier, ed. *Anthropologies of Modernity: Foucault, Governmentality and Life Politics*. Oxford: Blackwell, 2005.

Ipsen, Carl. *Dictating Demography: The Problem of Population in Fascist Italy*. Cambridge; New York: Cambridge University Press, 1996.

—. *Italy in the Age of Pinocchio: Children and Danger in the Liberal Era*. New York: Palgrave, 2006.

Israel, Giorgio, and Pietro Nastasi. *Scienza e razza nell'Italia fascista*. Bologna: Il mulino, 1998.

Iyob, Ruth. "Madamismo and Beyond: The Construction of Eritrean Women." *Nineteenth Century Contexts* 22.2 (2000): 217–238.

Jameson, Fredric. "Of Islands and Trenches: Naturalization and the Production of Utopian Discourse." *Diacritics* 7.2 (Summer 1977): 2–21.

—. *The Political Unconscious: Narrative as a Socially Symbolic Act*. Ithaca, NY: Cornell University Press, 1981.

Jay, Martin. *Downcast Eyes: The Denigration of Vision in Twentieth-Century French Thought*. Berkeley; Los Angeles, CA: University of California Press, 1994.

Jesi, Centro Furio, ed. *La menzogna della razza: documenti e immagini del razzismo e dell'antisemitismo fascista*. Bologna: Grafis, 1994.

Josz, Aurelia. "Fiammella francescana." *Nuova antologia,* March 16, 1913.

Kertzer, David I. "Gender Ideology and Infant Abandonment in Nineteenth-century Italy." *Journal of Interdisciplinary History* 22.1 (Summer 1991): 1–25.

Kjellén, Rudolf. *Staten som Lifsform.* Stockholm: Hugo Gebers, 1916.

Kornbluh, Anna. "Enjoying Law: Psychoanalysis and Sovereign Bodies." *JCRT* 12:1 (Spring 2012): 17–21.

Kracauer, Siegfried. *From Caligari to Hitler: A Psychological History of the German Film.* 1947. Princeton, NJ: Princeton University Press, 2004.

Labanca, Nicola. "'Un nero non può essere un bianco.' Il Museo Nazionale di Antropologia di Paolo Mantegazza e la colonia Eritrea." *L'Africa in vetrina. Storie di musei e di esposizioni coloniali in Italia.* Ed. Nicola Labanca. Treviso: Pagus, 1992. 69–106.

—. *Oltremare: storia dell'espansione coloniale italiana.* Bologna: Il mulino, 2002.

—. "The Embarrassment of Libya. History, Memory, and Politics in Contemporary Italy." *California Italian Studies Journal* 1.1 (2010): 1–18.

Lacan, Jacques. "The Mirror Stage as Formative of the I Function as Revealed in Psychoanalytic Experience." *Écrits.* Trans. Bruce Fink. New York: Norton, 2007. 75–82.

LaCapra, Dominick. *The Bounds of Race: Perspectives on Hegemony and Resistance.* Ithaca, NY: Cornell University Press, 1991.

—. "Trauma, Absence, Loss." *Critical Inquiry* 25:4 (Summer 1999): 696–727.

Lamarck, Jean Baptiste. *Philosophie zoologique, ou Exposition des considérations relatives à l'histoire naturelle des animaux.* Paris: F. Savy, 1873.

Lanaro, Silvio. *Nazione e lavoro. Saggio sulla cultura borghese in Italia.* Venice: Marsilio, 1979.

Landsberg, Alison. *Prosthetic Memory: The Transformation of American Remembrance in the Age of Mass Culture.* New York: Columbia University Press, 2004.

Landucci, Giovanni. *Darwinismo a Firenze: tra scienza e ideologia (1860–1900).* Florence: L.S. Olschki, 1977.

Landy, Marcia. *Fascism in Film: The Italian Commercial Cinema, 1931–1943.* Princeton, NJ: Princeton University Press, 1986.

Lant, Antonia. "Spazio per la razza in *Cabiria.*" *Cabiria e il suo tempo.* Turin: Museo Nazionale del Cinema, 1998. 212–222.

Larebo, Halie. *The Building of an Empire. Italian Land Policy and Practice in Ethiopia (1935–1941).* Oxford: Clarendon Press, 1994.

Lazzarato, Maurizio. "From Biopower to Biopolitics." www.generation-online.org/c/fcbiopolitics.htm (accessed July 6, 2014).

—. *Lavoro immateriale. Forme di vita e produzione di soggettività.* Verona: Ombre Corte, 1997.

Lazzati, Giulio. *Stormi d'Italia: storia dell'aviazione militare italiana.* Milan: Mursia, 1975.

Ledeen, Michael Arthur. *D'Annunzio: The First Duce*. New Brunswick, NJ: Transaction, 2002.
Lemke, Thomas. *Biopolitics: An Advanced Introduction*. New York: New York University, 2011.
Leopardi, Giacomo. *Canti*. Ed. C. Muscetta and G. Savoca. Turin: Einaudi, 1968.
—. *The Canti*. Trans. J.G. Nichols. New York: Routledge, 2003.
—. "Discorso sopra lo stato presente dei costumi degl'italiani." *Opere*, Vol. II. Milan: Rizzoli, 1823.
—. *Poesie e prose. Volume secondo*. Ed. Ronaldo Damiani. Milan: Mondadori, 2006. 441–480.
Lhamon, W.T. *Raising Cain: Blackface Performance from Jim Crow to Hip Hop*. Cambridge, MA: Harvard University Press, 1998.
Linke, Uli. *Blood and Nation: The European Aesthetics of Race*. Philadelphia, PA: University of Pennsylvania Press, 1999.
—. *German Bodies: Race and Representation after Hitler*. New York: Routledge, 1999.
"Lo 'scenario' di *Cabiria*." *Cabiria & Cabiria*. Ed. Silvio Alovisio and Alberto Barbera. Turin, 2006. 45–52.
Lombardi Satriani, Luigi "Realtà meridionale e conoscenza demologica. Linee per una storia degli studi demologici dagli anni postunitari alla conquista della Libia." *Studi antropologici italiani e rapporti di classe*. Ed. Sandra Puccini. Milan: Franco Angeli, 1980. 41–66.
Lombardi-Diop, Cristina. "Madre della nazione. Una donna italiana nell'Eritrea coloniale." *Africa Italia. Due continenti si avvicinano*. Ed. Stefano Bellucci and Sante Matteo. Santarcangelo di Romagna: Fara, 2001. 117–133.
—. "Postracial/Postcolonial Italy." *Postcolonial Italy: Challenging National Homogeneity*. Ed. Cristina Lombardi-Diop and Caterina Romeo. New York: Palgrave, 2012. 175–190.
—. "Writing the Female Frontier: Italian Women in Colonial Africa, 1880–1940." Ph.D. Dissertation, Comparative Literature. New York University, 1999.
Lombardi-Diop, Cristina, and Caterina Romeo, eds. *Postcolonial Italy: Challenging National Homogeneity*. New York: Palgrave, 2012.
Lombroso, Cesare. *Criminal Man*. Trans. Mary Gibson. Durham, NC: Duke Univeristy Press, 2006.
—. *Criminal Woman, the Prostitute, and the Normal Woman*. Trans. Mary Gibson. Durham, NC: Duke University Press, 2004.
—. *La donna delinquente, la prostituta e la donna normale*. Turin; Rome: L. Roux, 1893.
—. *L'uomo bianco e l'uomo di colore. Letture sull' origine e le varietà delle razze umane*. Padua: F. Sacchetto, 1871.
—. *L' uomo delinquente studiato in rapporto alla antropologia, alla medicina legale ed alle discipline carcerarie*. Milan: Ulrico-Hoelpi, 1876.

Lombroso, Cesare, and Giuseppe Pelaggi. *In Calabria, 1862–1897.* Sala Bolognese: A. Forni, 1988.

Lott, Eric. *Love and Theft: Blackface Minstrelsy and the American Working Class.* New York; Oxford: Oxford University Press, 1995.

—. "The Whiteness of Film Noir." *Whiteness: A Critical Reader.* Ed. Mike Hill. New York: New York University Press, 1997. 81–101.

Lowe, Lisa. "The Intimacies of Four Continents." *Haunted by Empire: Geographies of Intimacy in North American History.* Ed. Ann Laura Stoler. Durham, NC: Duke, 2006.

Lyttleton, A., ed. *Liberal and Fascist Italy.* Oxford; New York: Oxford University Press, 2002.

Malkiewicz, Kris. *Film Lighting: Talks with Hollywood's Cinematographers and Gaffers.* New York: Simon and Schuster, 1992.

Manenti, Carla, Nicolas Monti, and Giorgio Nicodemi, eds. *Luca Comerio fotografo e cineasta.* Milan: Electa, 1979.

Mannoni, Octave. "Je sais bien, mais quand-même ..." *Clefs pour l'imaginaire ou l'autre scène.* Paris: Editions du Seuil, 1969. 9–33.

Manovich, Lev. "Visual Technologies as Cognitive Prostheses: A Short History of the Externalization of the Mind." *The Prosthetic Impulse: From a Posthuman Present to a Biocultural Future.* Cambridge, MA: MIT Press, 2006. 203–219.

Mantegazza, Paolo. *Atlante della espressione del dolore.* Florence: Brogi, 1876.

—. *Fisiologia del dolore.* Florence: Felice Paggi, 1880.

—. *Fisiologia del piacere.* 1880. Milan: Bietti, 1947.

—. *Fisiologia dell'amore.* 1873. Florence: Bemporad, 1910.

—. *Fisiologia della donna.* Vol. II. Milan: Treves, 1893.

—. *Fisionomia e mimica.* Milan: Dumolard, 1881.

—. *Gli amori degli uomini.* Milan: P. Mantegazza, 1886.

—. *L'anno 3000. Sogno.* 1897. Bergamo: Lubrina, 1988.

—. *L'Anno 3000. Sogno di Paolo Mantegazza.* I rimossi. Milan: Lupetti, 2007.

—. *L'arte de prender moglie.* Milan: Treves, 1892.

—. *L'igiene dell'amore.* Milan: Brigola, 1879.

—. *Ricordi politici di un fantaccino del parlamento italiano.* Florence: R. Bemporad, 1897.

—. *Studi sulla etnologia dell'India.* Florence: Società Italiana d'Antropologia, 1886.

—. *Testa, libro per i giovinetti.* 1887. Milan: Fratelli Treves, 1921.

—. *The Year 3000. A Dream.* Trans. David Jacobson. Ed. Nicoletta Pireddu. Lincoln, NE: University of Nebraska Press, 2010.

—. *Un giorno a Madera. Una pagina dell'igiene d'amore.* 1868. Milan: G. Brigola, 1879.

—. *Un viaggio in Lapponia coll'amico Stefano Sommier.* Milan: Brigola, 1881.

Mantegazza, Paolo, and Neera. *Dizionario d'igiene per le famiglie.* 1881. Milan: Scheiwiller, 1985.

Mantovani, Claudia. "Per risanare il corpo della nazione: scelte individuali o responsabilità collettive?" *Rigenerare la società. L'eugenetica in Italia dalle origini ottocentesche agli anni Trenta*. Soveria Mannelli: Rubbettino, 2004. 87–114.

Manzoni, Alessandro. *Tutte le poesie*. Milan: Rizzoli, 1951.

Marcus, Millicent. *Italian Film in the Shadow of Auschwitz*. Toronto: Toronto University Press, 2007.

Marilotti, Gianni. *L'Italia e il Nord Africa. L'emigrazione sarda in Tunisia, 1848–1914*. Rome: Carocci, 2006.

Marinetti, F.T. *La battaglia di Tripoli (26 ottobre 1911)*. Padua: Tipografia Elzeviriana, 1912.

—. *Critical Writings*. Trans. Doug Thompson. Ed. Gunter Berghaus. New York: Farrar, Straus, Giroux, 2006.

—. *Futurismo e fascismo*. Foligno: Campitelli, 1924.

—. *Guerra sola igiene del mondo*. Milan: Edizioni Futuriste di Poesia, 1915.

—. *Mafarka il futurista*. 1910. Ed. L. Ballerini. Milan: Mondadori, 2003.

Marsella, Mauro. "Enrico Corradini's Italian Nationalism: The 'Right Wing' of the Fascist Synthesis." *Journal of Political Ideologies* 9.2 (June 2004): 203–224.

Martinelli, Vittorio. "Lasciate fare a noi, siamo forti." *Gli uomini forti*. Ed. Alberto Farassino and Tatti Sanguinetti. Milan: Mazzotta, 1983. 9–11.

Marx, Karl. *Capital, Volume 1*. London: Penguin, 1991.

Marx, Karl, and Frederick Engels, *Collected Works*. New York: International Publishers, 1975.

Masali, Luca. *I biplani di D'Annunzio. Romanzo*. Milan: Sironi Editore, 2006.

Masciola, Cristina. *Razza bastarda: romanzo*. Rome: Fanucci, 2007.

Matteo, Sante, ed. *ItaliAfrica: Bridging Continents and Cultures*. Stony Brook, NY: Forum Italicum, 2001.

Mbembe, Achille. *On the Postcolony*. Berkeley, CA: University of California Press, 2001.

McClintock, Anne. *Imperial Leather: Race, Gender, and Sexuality in the Colonial Contest*. New York: Routledge, 1995.

Melfa, Daniela. *Migrando a Sud. Coloni italiani in Tunisia, 1881–1939*. Rome: Aracne, 2008.

Mellino, Miguel. "De-Provincializing Italy: Notes on Race, Racialization, and Italy's Coloniality." *Postcolonial Italy: Challenging National Homogeneity*. Ed. Cristina Lombardi-Diop and Caterina Romeo. New York: Palgrave, 2012. 83–102.

Metz, Christian. *The Imaginary Signifier. Psychoanalysis and the Cinema*. Bloomington, IN: Indiana University Press, 1986.

Miele, Caterina. "Per un'archeologia del discorso razziale in Italia." *La razza al lavoro*. Ed. Anna Curcio and Miguel Mellio. Rome, 2012. 105–122.

Miller, Christopher L. *Blank Darkness: Africanist Discourse in French*. Chicago, IL: University of Chicago Press, 1985.

Ministero dell'Interno. Direzione Generale delle Carceri. *Statistica delle carceri, 1872*. Rome: Tipografia Artero, 1874.

Misek, Richard. *Chromatic Cinema: A History of Screen Color.* Malden, MA; Oxford: Blackwell, 2010.

Mitchell, Timothy. *Colonising Egypt*. Berkeley, CA: University of California Press, 1988.

Moe, Nelson. "The Geographical Poetics of Giovanni Verga." *The View from Vesuvius: Italian Culture and the Southern Question*. Berkeley, CA: University of California Press, 2002. 250–289.

—. *The View from Vesuvius: Italian Culture and the Southern Question*. Berkeley, CA: University of California Press, 2002.

Moliterno, Gino. *The A to Z of Italian Cinema (Historical Dictionary of Italian Cinema)*. Lanham, MD: Scarecrow Press, 2008.

Montag, Warren. "Toward a Conception of Racism without Race: Foucault and Contemporary Biopolitics." *PLI. The Warwick Journal of Philosophy* 13 (2002): 113–125.

Montagu, Ashley. *Man's Most Dangerous Myth: The Fallacy of Race*. 6th ed. Walnut Creek, CA: AltaMira Press, 1997.

—. *Statement on Race. An Annotated Elaboration and Exposition of the Four Statements on Race Issued by UNESCO*. Oxford: Oxford University Press, 1972.

—. *Statement on Race. An Extended Discussion in Plain Language of the UNESCO Statement by Experts on Race Problems*. New York: Schuman, 1951.

Moretti, Erica. "Pedagogia della terra. Gli esperimenti educative di Alice Hallgarten e Leopoldo Franchetti." *Zapruder. Storie di classe*. 21 (2011): 140–145.

Morrison, Toni. *Playing in the Dark: Whiteness and the Literary Imagination*. Cambridge, MA: Harvard University Press, 1992.

Morton, Peter. *The Vital Science: Biology and the Literary Imagination 1860–1900*. London: George Allen & Unwin, 1984.

Mosse, George. *Nationalism and Sexuality: Middle-class Morality and Sexual Norms in Modern Europe*. Madison, WI: University of Wisconsin Press, 1988.

—. "The Poet and the Exercise of Political Power: Gabriele D'Annunzio." *Yearbook of Comparative and General Literature* 22 (1973): 32–41.

Mussolini, Benito. "Circolare ai prefetti, 10 [?] agosto, 1927." *Opera omnia*. Ed. E and D. Sumsel. Florence: La Fenice, 1957.

Musumeci, Mario, and Sergio Toffetti, eds. *Da* La presa di Roma *a* Il piccolo garibaldino. *Risorgimento, massoneria e istituzioni: l'immagine della Nazione nel cinema muto (1905–1909)*. Rome: Gangemi Editore, 2007.

Negash, Tekeste. *Italian Colonialism in Eritrea, 1882–1941: Policies, Praxis and Impact*. Uppsala; Stockholm: Universitatis Upsaliensis, 1987.

Niceforo, Alfredo. *L'Italia barbara contemporanea*. Milan: Remo Sandron, 1898.

Nye, Robert A. "The Rise and Fall of the Eugenics Empire: Recent Perspectives on the Impact of Biomedical Thought in Modern Society." *The Historical Journal* 36.3 (September 1993): 687–700.

O'Healy, Aine. "Screening Intimacy and Racial Difference in Postcolonial Italy." *Postcolonial Italy: Challenging National Homogeneity*. Ed. Cristina Lombardi-Diop and Caterina Romeo. New York: Palgrave, 2012. 205–220.

Oriani, Alfredo. *Fino a Dogali*. Bari: Laterza, 1918.

—. *Opera Omnia di Alfredo Oriani*. Ed. Benito Mussolini. Bologna: Cappelli, 1924–1933.

—. *L'ora d'Africa*. Bologna: Cappelli, 1940.

Palma, Silvana. *L'Italia coloniale*. Rome: Riuniti, 1999.

—. "The Seen, The Unseen, The Invented: Misrepresentations of African 'Otherness' in the Making of a Colony. Eritrea 1885–1896." *Cahiers d'Études Africaines* 45.177 (2005): 39–69.

Palumbo, Patrizia, ed. *A Place in the Sun: Africa in Italian Colonial Culture from Post-unification to the Present*. Berkeley, CA: University of California Press, 2003.

Papini, Giovanni. *Passato remoto, 1885–1914*. Florence: L'Arco, 1948.

Parati, Graziella. *Migration Italy: The Art of Talking Back in a Destination Culture*. Toronto; Buffalo: University of Toronto Press, 2005.

Pascoli, Giovanni. *La grande proletaria si è mossa*. Bologna: Zanichelli, 1911.

—. *Limpido rivo*. Bologna: Zanchelli, 1913.

Pasini, Mirella. "Fisionomia e mimica. Paolo Mantegazza interprete di Darwin." *Oltre lo sguardo: la fisiognomica e lo studio della natura umana* (Milan: Mimesis, 2011) 127–140.

Pasini, Walter. *Paolo Mantegazza, ovvero l'elogio dell'eclettismo*. Rimini: Panozzo, 1999.

Passerini, Luisa. *Fascism in Popular Memory: The Cultural Experience of the Turin Working Class*. Cambridge: Cambridge University Press, 1987.

—. "Memories of Resistance, Resistances of Memory." *European Memories of the Second World War*. Ed. Hemlut Peitsch. New York: Berghahn Books, 1999. 288–296.

—. *Women Migrants from East to West: Gender, Mobility and Belonging in Contemporary Europe*. New York: Berghahn Books, 2007.

Patriarca, Silvana. "Indolence and Regeneration: Tropes and Tensions of Risorigmento Patriotism." *American Historical Review* 110 (2005): 380–408.

—. *Italian Vices: Nation and Character from the Risorgimento to the Republic*. Cambridge; New York: Cambridge University Press, 2010.

—. *Numbers and Nationhood: Writing Statistics in Nineteenth-century Italy*. Cambridge: Cambridge University Press, 1996.

Pelizzari, Maria Antonella. *Percorsi della fotografia in Italia*. Rome: Contrasto, 2011.

—. *Photography and Italy*. London: Reaktion Books, 2011.

Pender, E.E. "Spiritual Pregnancy in Plato's Symposium." *Classical Quarterly* 42 (1992): 72–86.

Pendola, Marinette. *Gli italiani di Tunisia. Storia di una comunità (XIX–XX secolo)*. Perugia: Editoriale Umbra, 2007.

Persano, Enrico. "L'igiene coloniale." *L'Agricoltura coloniale* I.I (Jan–Feb 1908): 120–124.

Petrarca, Francesco. *The Canzoniere, or Rerum vulgarium fragmenta*. Trans. Mark Musa. Bloomington, IN: Indiana University Press, 1996.

—. *Rerum vulgarium fragmenta*. Ed. M. Santagata. Milan: Mondadori, 1996.

Pezzino, Paolo. *Leopoldo e Alice Franchetti e il loro tempo*. Città di Castello: Petruzzi, 2002.

Piaggia, Carlo. *Nella terra dei Niam-Niam (1863–1865)*. Ed. Ezio Bassani. Lucca: Maria Pacini Fazzi, 1978.

Pianavia Vivaldi, Rosalia. *Tre anni in Eritrea*. Milan: Cogliati, 1901.

Pickering, Michael. *Blackface Minstrelsy in Britain*. Aldershot, UK; Burlington, VT: Ashgate, 2008.

Pietz, William. "The Problem of the Fetish, I." *RES: Anthropology and Aesthetics* 9 (Spring 1985): 5–17.

Pinkus, Karen. *Bodily Regimes: Italian Advertising under Fascism*. Minneapolis, MN: University of Minnesota Press, 1995.

Pireddu, Nicoletta. *Antropologi alla corte della bellezza: decadenza ed economia simbolica nell'Europa fin de siècle*. Verona: Fiorini, 2002.

—. "The Anthropological Roots of Italian Cultural Studies." *Italian Cultural Studies* 29 (2001): 66–88.

—. "Gabriele D'Annunzio. The Art of Squandering and the Economy of Sacrifice." *The Question of the Gift: Essays across Disciplines*. Ed. Mark Osteen. New York: Routledge, 2002. 172–190.

—. "Introduction. Paolo Mantegazza, Fabulator of the Future." *The Year 3000*. Trans. David Jacobson. Ed. Nicoletta Pireddu. Lincoln, NE; London: University of Nebraska Press, 2010. 1–56.

—. "Introduction. Paolo Mantegazza: A Scientist and His Ecstasies." *Paolo Mantegazza. The Physiology of Love and Other Writings*. Trans. David Jacobson. Ed. Nicoletta Pireddu. Toronto: University of Toronto Press, 2007. 3–53.

—, ed. *Paolo Mantegazza. The Physiology of Love and Other Writings*. Toronto: University of Toronto Press, 2007.

Pisanty, Valentina, and Luca Bonafé. *Educare all'odio: La difesa della razza, 1938–1943*. Milan: Motta On Line, 2003.

Ponzanesi, Sandra. "The Color of Love: Madamismo and Interracial Relationships in the Italian Colonies." *Research in African Literatures* 43.2 (Summer 2012): 155–172.

Portelli, Alessanro. *L'ordine è già stato eseguito. Roma, le Fosse Ardeatine, la memoria*. Rome: Donizelli, 1999.

—. *The Order Has Been Carried Out: History, Memory, and Meaning of a Nazi Massacre in Rome*. Basingstoke: Palgrave, 2003.

Pratt, Mary Louise. *Imperial Eyes: Travel Writing and Transculturation*. New York: Routledge, 1992.

Prolo, Maria Adriana. *Storia del cinema muto italiano*. Milan: Poligono, 1951.

Puccini, Sandra. "Giustiniano Nicolucci e gli studi demo-etno-antropologici italiani dell'Ottocento." *La ricerca folklorica* 12 (October 1985): 131–135.

—. *L'Uomo e gli uomini: scritti di antropologi italiani dell'Ottocento*. Rome: Centro Informazione Stampa Universitaria, 1991.

—. "I viaggi di Paolo Mantegazza. Tra divulgazione, letteratura e antropologia." *Paolo Mantegazza. Medico, antropologo, viaggiatore*. Ed. Cosimo Chiarelli and Walter Pasini. Florence: Firenze University Press, 2002. 49–74.

Pushed Back, Pushed Around. Italy's Forced Return of Boat Migrants and Asylum Seekers, Libya's Mistreatment of Migrants and Asylum Seekers. New York: Human Rights Watch, 2009.

Quine, Maria Sophia. "Making Italians: Aryanism and Anthropology in Italy during the Risorgimento." *Crafting Humans: From Genesis to Eugenics and Beyond*. Ed. Marius Turda. Taipei: National University of Taiwan Press, 2013. 127–152.

Rabinow, Paul. *French Modern: Norms and Forms of the Social Environment*. Cambridge, MA: MIT Press, 1989.

Rabinow, Paul, and Nikolas Rose, eds. *The Essential Foucault: Selections from the Essential Works of Foucault. 1954–1974*. New York: New Press, 2003.

Raimondi, Ezio, and Gabriella Fenocchio. *La letteratura italiana: il Novecento*. Milan: Mondadori, 2004.

Rainero, Romain. *L'Italia e il Nordafrica contemporaneo*. Milan: Mazorati, 1988.

—. *Les Italiens dans la Tunisie contemporaine*. Paris: Publisud, 2002.

—. *I primi tentativi di colonizzazione agricola e di popolamento dell'Eritrea, 1890–1895*. Milan: Marzorati, 1960.

Re, Lucia. "Gabriele D'Annunzio's novel *Le vergini delle rocce:* 'Una cosa naturale vista in un grande specchio'." *Stanford Italian Review* III (Fall 1983): 241–272.

—. "Gabriele D'Annunzio's Theater of Memory: Il Vittoriale degli Italiani." *DAPA* (Winter 1987): 6–50.

—. "Italians and the Invention of Race: The Poetics and Politics of Difference in the Struggle over Libya, 1890–1913." *California Italian Studies Journal* 1.1 (2010): 1–58.

Redi, Riccardo. *Cinema italiano sotto il fascismo*. Venice: Marsilio, 1979.

Reich, Jacqueline. "Slave to Fashion: Masculinity, Suits, and the Maciste Films of Italian Silent Cinema." *Fashion in Film*. Ed. Adrienne Munich. Bloomington, IN: Indiana University Press, 2011. 236–259.

Reich, Jacqueline, and Piero Garofalo, eds. *Re-viewing Fascism: Italian Cinema, 1922–1943*. Bloomington, IN: Indiana University Press, 2002.

Renzi, Renzo. *Il fascismo involontario e altri scritti*. Bologna: Cappelli, 1975.

Rhodes, John David. "'Our Beautiful and Glorious Lives': The Rhetoric of Nationalism in Early Italian Film Periodicals." *Film History* 12 (2000): 308–321.

Ricci, Laura. *La lingua dell'impero. Comunicazione, letteratura e propaganda nell'età del colonialismo italiano*. Rome: Carocci, 2005.

Ricci, Steven. *Cinema and Fascism: Italian Film and Society, 1922–1943*. Berkeley, CA: University of California Press, 2008.

Righi, Andrea. *Biopolitics and Social Change in Italy: From Gramsci to Pasolini to Negri*. New York: Palgrave Macmillan, 2011.

Rochat, Giorgio. "Il colonialismo italiano." *L'Africa in vetrina. Storie di musei e di esposizioni coloniali in Italia*. Ed. Nicola Labanca. Treviso: Pagus, 1992. 9–15.

—. *Le guerre italiane in Libia e in Etiopia dal 1896 al 1939*. Treviso: Pagus, 1991.

Roda, Vittorio. *I fantasmi della ragione: fantastico, scienza e fantascienza nella letteratura italiana fra Otto e Novecento*. Naples: Liguori, 1996.

Rogin, Michael Paul. *Blackface, White Noise: Jewish Immigrants in the Hollywood Melting Pot*. Berkeley, CA: University of California Press, 1996.

—. "The Sword Became a Flashing Vision." *'Ronald Reagan,' The Movie, and Other Episodes in Political Demonology*. Berkeley; Los Angeles, CA: University of California Press, 1987. 190–235.

Romano, Ruggiero, and Corrado Vivanti, eds. *Storia d'Italia: Dall'Unità a oggi*. Vol. IV. Turin: Einaudi, 1976.

Romano, Sergio. *Le Italie parallele. Perché l'Italia non riesce a diventare un paese moderno*. Milan: Longanesi, 1996.

Rosengarten, Frank. "Homo Siculus: Essentialism in the Writing of Giovanni Verga, Giuseppe Tomasi di Lampedusa, and Leonardo Sciascia." *Italy's "Southern Question": Orientalism in One Country*. Ed. Jane Schneider. Oxford; New York: Berg, 1998.

Rusconi, Gian Enrico. *Se cessiamo di essere una nazione*. Bologna: Mulino, 1996.

Rushing, Robert. "Gentlemen Prefer Hercules: Desire/ Identification/ Beefcake." *Camera Obscura* 23.3 (2008): 159–192.

Russo, Luigi. *I narratori*. Milan: Principato, 1958.

Said, Edward. *Culture and Imperialism*. New York: Knopf, 1993.

—. *Orientalism*. New York: Pantheon, 1978.

Salaris, Claudia. *Alla festa della rivoluzione: artisti e libertari con D'Annunzio a Fiume*. Bologna: Il mulino, 2002.

Salinari, Carlo. *Miti e coscienza del decadentismo italiano: D'Annunzio, Pascoli, Fogazzaro e Pirandello*. Milan: Feltrinelli, 1960.

Salvadori, Massimo. *Il mito del buongoverno. La questione meridionale da Cavour a Gramsci*. Turin: Einaudi, 1963.

Santner, Eric. *The Royal Remains: The People's Two Bodies and the Endgames of Sovereignty*. Chicago, IL: University of Chicago Press, 2012.
Sartini Blum, Cinzia. "The Futurist Refashioning of the Universe." *South Central Review* 13.2–3 (Summer–Fall 1996): 82–104.
Scarlini, Luca. *D'Annunzio a Little Italy. Le avventure del Vate nel mondo dell'emigrazione*. Rome: Donzelli, 2008.
Schenk, Irmbert. "The Cinematic Support to National(istic) Mythology: The Italian Peplum 1910–1930." *Globalization, Cultural Identities, and Media Representations*. Ed. Natascha Gentz and Stefan Kramer. Albany, NY: State University of New York Press. 153–167.
Schiavone, Aldo. *Italiani senza Italia: Storia e identità*. Turin: Einaudi, 1998.
Schnapp, Jeffrey. "Nietzsche's Italian Style." *Stanford Italian Review: Nietzsche in Italy* (1986): 247–264.
—. "Propeller Talk." *Modernism/Modernity* 1.3 (1994).
—. "The Fabric of Modern Times." *Critical Inquiry* 24.1 (Autumn 1997): 191–245.
Schneider, Jane, ed. *Italy's "Southern Question": Orientalism in One Country*. Oxford; New York: Berg, 1998.
Schram-Pighi, Laura. *La narrativa italiana di utopia dal 1750 al 1915*. Ravenna: Longo, 2003.
Schwartz, Vanessa. "Film and History." *The SAGE Handbook of Film Studies*. Ed. James Donald and Michael Renov. Thousand Oaks, CA: SAGE Publications, 2008. 199–215.
"-scope, *comb. form*." *The Oxford English Dictionary Online*. 2 ed: Oxford University Press, 2011.
"scope, n. 2a." *The Oxford English Dictionary Online*. 2 ed: Oxford University Press, 2011.
Scott, David. "Colonial Governmentality." *Anthropologies of Modernity: Foucault, Governmentality, and Life Politics*. Ed. Jonathan Xavier Inda. Oxford: Blackwell, 2005. 23–49.
Sekula, Allan. "The Body and the Archive." *October* 39 (1986) 3–64.
Sergi, Giuseppe. *The Mediterranean Race: A Study of the Origin of European Peoples*. 1895. Oosterhout, Netherlands: Anthropological Publications, 1967.
—. *Africa. Antropologia della stirpe camitica (specie eurafricana)*. Turin: Fratelli Bocca Editori, 1897.
Seton-Watson, Christopher. "Italy's Imperial Hangover." *Journal of Contemporary History* 15.1 (1980): 169–179.
Shukin, Nicole. *Animal Capital: Rendering Life in Biopolitical Times*. Minneapolis, MN: University of Minnesota Press, 2009.
Snowden, Frank M. *The Conquest of Malaria: Italy, 1900–1962*. New Haven, CT: Yale University Press, 2006.
—. *Naples in the Time of Cholera, 1884–1911*. Cambridge; New York: Cambridge University Press, 1995.

Sòrgoni, Barbara. "Italian Anthropology and The Africans: The Early Colonial Period." *A Place in the Sun: Africa in Italian Colonial Culture from Post-Unification to the Present*. Ed. Patrizia Palumbo. Berkeley, CA: University of California Press, 2003. 62–80.

—. *Parole e corpi. Antropologia, discorso giuridico e politiche sessuali interraziali nella colonia Eritrea: 1890–1941*. Naples: Liguori, 1998.

Spackman, Barbara. *Decadent Genealogies: The Rhetoric of Sickness from Baudelaire to D'Annunzio*. Ithaca, NY: Cornell University Press, 1989.

—. *Fascist Virilities: Rhetoric, Ideology, and Social Fantasy in Italy*. Minneapolis, MN: University of Minnesota Press, 1996.

—. "*Il verbo (e)sangue:* Gabriele D'Annunzio and the Ritualization of Violence." *Quaderni d'Italianistica* IV.2 (1983): 218–229.

Speciale, Giuseppe. *Giudici e razza nell'Italia fascista*. Turin: G. Giappichelli, 2007.

Spinosa, Antonio, and Francesco Perfetti. *Mussolini: razzista riluttante*. Rome: Bonacci, 1994.

Stepan, Nancy Leys. "Race, Gender, Science and Citizenship." *Gender and History* 10.1 (1998): 26–52.

Stewart-Steinberg, Suzanne. *The Pinocchio Effect: On Making Italians (1860–1920)*. Chicago, IL: University of Chicago Press, 2007.

Stocking, George. "Objects and Others: Essays on Museums and Material Culture." *History of Anthropology*, Vol. III. Ed. George Stocking. Madison, WI: University of Wisconsin Press, 1985.

—. *Victorian Anthropology*. New York: Free Press, 1987.

Stoler, Ann Laura. *Carnal Knowledge and Imperial Power: Race and the Intimate in Colonial Rule*. Berkeley, CA: University of California Press, 2002.

—. *Race and the Education of Desire: Foucault's History of Sexuality and the Colonial Order of Things*. Durham, NC: Duke University Press, 1995.

Strausbaugh, John. *Black Like You: Blackface, Whiteface, Insult, and Imitation in American Popular Culture*. New York: Penguin, 2006.

Syrimis, Michael. "Decadent Repititions, Technological Variations in Gabriele D'Annunzio's *Forse che sì forse che no*." *Romance Notes* 45.2 (2005): 141–150.

Taddia, Irma. *L'eritrea colonia 1890–1952: paesaggi, struttura, uomini del colonialismo*. Milan: Agnelli, 1986.

Taylor, P.M. "Paolo Mantegazza (1831–1910): Reassessing the Florentine School of Anthropology in Pre-Fascist Italy (up to 1925)." *Antropologia contemporanea* 10:1–2 (1987): 1–16.

Teti, Vito. *La razza maledetta: origini del pregiudizio antimeridionale*. Roma: Manifestolibri, 1993.

Tomasello, Giovanna. *L'Africa tra mito e realtà: storia della letteratura coloniale italiana*. Palermo: Sellerio, 2004.

—. *Letteratura coloniale italiana dalle avanguardie al fascismo*. Palermo: Sellerio, 1984.

Tomasi di Lampedusa, Giuseppe. *Il gattopardo*. Milan: Feltrinelli, 1957.

Trento, Giovanna. "From Marinetti to Pasolini: Massawa, the Red Sea, and the Construction of 'Mediterranean Africa' in Italian Literature and Cinema." *Northeast African Studies* 12.1 (2012): 273–307.

Triulzi, Alessandro. "Displacing the Event: Hybrid Memories of Postcolonial Italy." *National Belongings. Hybridity in Italian Colonial and Postcolonial Cultures*. Oxford; Bern: Peter Lang, 2010. 23–40.

Tunander, Ola. "Swedish-German Politics for a New Century. Rudolf Kjellén's 'The State as a Living Organism'." *Review of International Studies* 27.3 (2001): 451–463.

Valesio, Paolo, ed. *D'Annunzio a Yale: atti del convegno*. Gardone Riviera: Fondazione il Vittoriale degli italiani, 1989.

—. *Gabriele D'Annunzio: The Dark Flame*. New Haven, CT; London: Yale University Press, 1992.

Van den Abbeele, Georges. *Travel as Metaphor*. Minneapolis, MN: University of Minnesota, 1992.

Verdicchio, Pasquale. *Bound by Distance: Rethinking Nationalism Through the Italian Diaspora*. Madison, WI: Fairleigh Dickinson University Press, 1997.

—. "Introduction." *Antonio Gramsci. The Southern Question*. Ed. Pasquale Verdicchio. West Lafayette, IN: Bordighera, 1995.

Vettori, Vittorio. *Gabriele D'Annunzio e il mito del superuomo*. Castello di Borgo alla Collina (Arezzo): Accademia casentinese di lettere, arti, scienze ed economia, 1981.

Villari, Pasquale. *Le lettere meridionali ed altri scritti sulla questione sociale in Italia*. Naples: Guida, 1979.

Virilio, Paul. *War and Cinema. The Logistics of Perception*. Brooklyn, NY: Verso, 1989.

Virno, Paolo. *A Grammar of the Multitude: For an Analysis of Contemporary Forms of Life*. New York: Semiotext(e), 2004.

Von Clausewitz, Carl. *On War*. London: K. Paul, Trench, Trubner, 1908.

Von Henneberg, Krystyna Clara. "Monuments, Public Space, and the Memory of Empire in Modern Italy." *History and Memory* 16.1 (2004): 37–85.

Waldbaum, Roberta. "Alice Hallgarten Franchetti. Cultivating Learning Landscapes." *The Grand Tour Lives On*. Bologna: Cooperativa Libraria Universitaria Editrice Bologna (CLUEB), 2006. 119–135.

Welch, Rhiannon Noel. "Intimate Truth and (Post)colonial Knowledge in Shirin Ramzanali Fazel's *Lontano da Mogadiscio*." *National Belongings: Hybridity in Italian Colonial and Postcolonial Cultures*. Ed. Jacqueline Andall and Derek Duncan. Bern: Peter Lang, 2010. 215–233.

—. "Razza e (ri)produttività. Per una lettura biopolitica della razza nell'Italia postunitaria e contemporanea." *L'Italia postcoloniale*. Ed. Cristina Lombardi-Diop and Caterina Romeo. Florence: Le Monnier, 2014. 79–91.

—. "Race and Colonial (Re)productivity." *Annali d'italianistica* 32 (2014). *From Otium & Occupatio to Work & Labor in Italian Culture.* Ed. Norma Bouchard and Valerio Ferme. 197–214.

Williams, Linda. *Playing the Race Card.* Princeton, NJ: Princeton University Press, 2002.

Winston, Brian. *Technologies of Seeing: Photography, Cinematography, and Television.* London: British Film Institute, 1996.

Wittman, Laura. "Mystical Insight and Psychology in the Fin-de-Siècle Novel: Huysmans, Bourget, D'Annunzio." *Forum Italicum* 42 (2012): 30–51.

—. *The Tomb of the Unknown Soldier, Modern Mourning, and the Reinvention of the Mystical Body.* Toronto: University of Toronto Press, 2011.

Wohl, Robert. *The Spectacle of Flight: Aviation and the Western Imagination, 1920–1950.* New Haven, CT: Yale University Press, 2005.

Wong, Aliza S. *Race and the Nation in Liberal Italy, 1861–1911: Meridionalism, Empire, and Diaspora.* New York: Palgrave Macmillan, 2006.

Woodhouse, John Robert. *Gabriele D'Annunzio: Defiant Archangel.* Oxford: Oxford University Press, 1998.

Yegenoglu, Medya. *Colonial Fantasies: Toward a Feminist Reading of Orientalism.* Cambridge: Cambridge University Press, 1998.

Young, Robert J.C. *Colonial Desire: Hybridity in Theory, Culture and Race.* London: Routledge, 1995.

—. "Foucault on Race and Colonialism." *New Formations. Michel Foucault: J'accuse* 25 (Summer 1995): 57–65.

—. *Postcolonialism: An Historical Introduction.* Oxford: Blackwell, 2001.

Zantop, Susanne. *Colonial Fantasies: Conquest, Family and Nation in Pre-colonial Germany, 1770–1870.* Durham, NC: Duke University Press, 1997.

Zarrilli, Ludovica. "Il Darwin della foto. Smorfie e clic per studiare l'uomo." *Corriere fiorentino.* October 8, 2010: 12.

Žižek, Slavoj. *In Defense of Lost Causes.* New York: Verso, 2008.

—. *The Sublime Object of Ideology.* New York: Verso, 1989.

Zuberi, Tukufu. *Thicker than Blood: How Racial Statistics Lie.* Minneapolis, MN: University of Minnesota Press, 2003.

Index